THE FUTURE OF EVERYTHING

TIM DUNLOP is a writer, academic and popular speaker. Author of *Why the Future is Workless*, he has a PhD in political philosophy, and has written and broadcast extensively on US and Australian politics, the media and the future of work in *The Guardian* and elsewhere. He lives in Melbourne and tweets @timdunlop.

For my Mum

THE FUTURE OF EVERYTHING

Big, Audacious Ideas for a Better World

TIM DUNLOP

NEWSOUTH

A NewSouth book

Published by
NewSouth Publishing
University of New South Wales Press Ltd
University of New South Wales
Sydney NSW 2052
AUSTRALIA
newsouthpublishing.com

First published 2018

10 9 8 7 6 5 4 3 2 1

ISBN: 9781742235646 (paperback)
9781742244327 (ebook)
9781742248752 (ePDF)

A catalogue record for this book is available from the National Library of Australia

Design Josephine Pajor-Markus
Cover design Luke Causby, Blue Cork
Cover image BAIVECTOR/Adobe Stock
Printer Griffin Press

This book is printed on paper using fibre supplied from plantation or sustainably managed forests.

CONTENTS

PROLOGUE

Those of us who want a better, fairer world need to stop playing defence and start changing the ground on which decisions about our lives are made. We need to stop arguing about how to force those who have so much to give a little back to the rest of us and start demanding that everyone get a fair share of the wealth we all help create.

Enough with sensible centrism, with polite and nervous requests – like so many Olivers holding out our bowls for another spoonful of slop – and a bit more of the impertinent demand. Let us rethink from the ground up our ideas about government, work, education and wealth distribution, so that the world we live in, in the first instance, *survives*, and then is managed in ways that benefit the many and not just the few.

We are in the midst of the greatest technological revolution of all time, one that is likely to put the necessities as well as the luxuries of life within increasingly easy grasp of us all, while also making possible new ways of living, ways that are more cooperative and that better lead to shared outcomes. But such advances will be for nothing if we continue as we are, concentrating wealth in the hands of the few, using this technology for control and surveillance rather than liberation and fairness.

Technology, then, is part of the solution, but it is quite literally just a tool, a means to help us achieve a better life. We should use it to make our economies more efficient and productive, but only in the service of the higher purpose of a decent life for all, not as an end in itself.

To achieve that, we need to organise and wield power so that we have:

- more everyday people in government deciding the course of our future
- more workplaces owned and controlled by the people who work in them
- essential public assets such as water and power publicly owned
- a guaranteed basic income so people have the time and financial security to live the life they choose
- a media that better reports the truth and citizens who can recognise it
- an education system that prepares us for the future, not the past
- joy in all its complexity.

Let's cut the cake more fairly so we can stop arguing over the crumbs.

INTRODUCTION

> Young people, people this society blatantly short-changes and betrays, are looking for intelligent, realistic, long-term thinking: not another ranting ideology, but a practical working hypothesis, a methodology of how to regain control of where we're going. Achieving that control will require a revolution as powerful, as deeply affecting society as a whole, as the force it wants to harness.
>
> *Ursula K Le Guin*

Democracy doesn't work without a reasonable degree of economic equality, a certain level of financial security for everyone, a functioning free press, the ability to influence the way in which we are governed and a reliable amount of free time so that we can enjoy the life we live together. Until recently, in developed nations, most of us could rely on having a job that pays a living wage, an education system that prepares us for that work as well as for citizenship, a representative system of government that operates in all of our interests, and a media that tells us the truth about what is going on. That framework is falling apart. Some of it can repaired. Some of it needs to be replaced.

We have become incredibly good at analysing where our society has gone wrong. There are far fewer attempts to explain how we might fix it.

What you will find in this book is a comprehensive set of changes that, I argue, will make the world a better place: fairer, more democratic, less violent, more joyous. It is an audacious agenda for real democratic change – in work, wealth, journalism, government, education and the natural world – to recognise we are all in this together and that to change anything we must take back control from those who are currently failing us.

Welcome to the future of everything …

*

In 2016, Swedish billionaire László Szombatfalvy offered US$5 million to anyone who could come up with a scheme for fixing the world, or, in the words of the official announcement, for re-envisioning global governance:

> We are delighted to announce the launch of the Global Challenges Prize 2017: A New Shape. This competition is a quest to find new models of global cooperation capable of handling global risks. It will award US$5 million in prizes for the best ideas that re-envision global governance for the 21st century. Be part of the effort to safeguard our world for future generations. Be part of the global conversation. Help to change the shape of things to come.

Offering a large sum of money to inspire people to think through the big problems confronting the world sounds like a reasonable, and attractive, idea. I mean, that's a hell of lot more than you would get for, say, a book on the same topic. And who knows, maybe such a competition might just inspire some good ideas we can all benefit from. You certainly can't fault the guy for trying, and even if you happened to be a bit cynical about the chances of such a competition coming up with a workable scheme, what possible harm could it do? If some rich guy wants to give away a truckload of money for a well-meaning competition, well then, knock yourself out.

The trouble is, such a competition is not harmless. Not only is it unlikely to achieve its stated goals, it indulges in an approach to 'global governance' that works against the very thing it is trying to achieve. It cultivates the fantasy that if we can just find the right financial incentive, the answers will present themselves. It is a species of that immense bullshit we generally call neoliberalism, a market-based understanding of society whose promise of a better life died a death in the global financial crisis of 2007, even if its advocates are yet to notice. The New Shape prize embodies an individualistic and capital-based approach to what is actually a communal problem. Matters of global governance are not going to be solved by people running off to their rooms and dreaming up theoretical solutions to the wicked problems that confront us, but by coming together in school rooms and town halls and parliaments and engaging in the difficult, thankless and very unsexy work of public argument. To accept the terms of a prize like this is to be defeated before you even start.

Silicon Valley is full of entrepreneurs who share this fix-the-world fantasy. Throw enough money at it, get enough smart tech people together, find the right incentives, remove the friction from the customer experience, and every problem will ultimately yield. This works fine if your goal is to build a search engine, or a platform for staying in contact with your friends and relatives, or maybe even for building a driverless car. But it doesn't work for governance, as a look at the track record of many tech companies will show, with their ongoing failures to confront basic ethical issues around gender, race and equality within their own ranks. Social critic and web developer Maciej Cegłowski makes a similar point in his essay 'Notes from an emergency':

> Real problems are messy. Tech culture prefers to solve harder, more abstract problems that haven't been sullied by contact with reality. So they worry about how to give Mars an earth-like climate, rather than how to give Earth an earth-like climate. They debate how to make a morally benevolent God-like AI, rather than figuring out how to put ethical guard rails around the more pedestrian AI they are introducing into every area of people's lives.

We need to get this through our heads: technology is not going to save us. A competition about new forms of global governance is not going to save us. Outsmarting someone on Twitter is not going to save us, nor is pointing out that President X or Prime Minister Y is a hypocrite. Simply knowing that you are being wronged, being able to point to examples of it, even being able to document the centuries-long history of your oppression, is not enough if the structural prejudice persists. There are no shortcuts here. As ever, there is only politics, the willingness to fight and organise and win people over to your plans on a mutually agreed basis of compromise and tolerance. For people of the left, political progressives who believe in the common good, there is no substitute for allowing ordinary people to be the instrument of their own betterment, and this is the idea at the heart of this book: all the reforms suggested in these pages begin with the idea that the people themselves should be in control of their own lives, and they should be given the means to achieve that.

Since publishing my book *Why the Future is Workless* in 2016, I've talked to people at a range of events about how our lives are likely to change. At ideas festivals, in radio interviews, at conferences, and in pubs and bookshops, I've heard people raise their concerns. I've also spoken with experts in everything from artificial intelligence (AI) to education, with union leaders, business leaders and politicians. I wouldn't say there is a uniform view among these disparate groups, but they do all realise something big is happening in the world, that a significant part of it relates to work and technology, and that we had better get out in front of it before we are overwhelmed. *Workless* was partly about the nature of new technologies that are getting smarter and more readily available. But it was also about the social and political ramifications of a postwork world, and these have dominated most discussions I've had since. What is that world going to look like? How will we live in it?

To begin to answer these questions, we need some context. You see, a lot of people – many of them economists – insist that concern about technological unemployment is exaggerated. They claim that new technology always creates as many jobs as it destroys and we therefore have nothing to worry about, that the market will take care of everything. It's the same tired neoliberal assertion that we have been hearing for the last 40 years. It gave us the global financial crisis, austerity, the rustbelt, and ultimately Brexit and Trump.

This sort of explanation is deaf to politics and social matters, presumes that markets are naturally occurring, operate in the general interest and are best facilitated without any sort of government interference, where 'interference' includes everything from setting a minimum wage to safety regulations for buildings. This attitude is summed up in the words of former British prime minister David Cameron, who in 2012, the year Britain hosted the Olympic Games, announced that his new year's resolution was to 'kill off the health and safety culture for good'. He said that this

> is not something government can do alone. It needs a change in the national mindset. We need to realise, collectively, that we cannot eliminate risk and that some accidents are inevitable. We need to take responsibility for our actions and rely on common sense rather than procedure. Above all, we need to give British businesses the freedom and discretion they need to grow, create jobs and drive our economy forward.

Cameron's dream was perfectly realised in June 2017 when Grenfell Tower, a block of flats for low-income people in the posh neighbourhood of Kensington and Chelsea, burned to the ground, killing 71 people and injuring 70 more. Residents had been complaining for years that basic health and safety features were missing from the building, but taking advantage of Cameron's desire to 'kill off the health and safety culture for good', the various authorities involved had ignored such pleas, while the building itself had been clad in an outer covering that was known to be combustible but was chosen because it was cheaper than safer materials.

When basic systems of governance not only fail but are actively undermined by politicians more interested in making the world safe for profits than for people, is it any wonder people look to the extremes for answers?

In the wake of Donald Trump's installation as US president and the success of far-right parties in Europe, the mainstream press has filled up with earnest op-eds from journalists and commentators urging us to embrace bipartisanship – what they often call the 'sensible centre' – and to resist the extremes of left and right. As well-meaning as such calls might be, they are nonsense, not least because there is no such thing as a political centre, sensible or otherwise. Yes, we can place most policies on a spectrum between a notional 'left' and 'right', and yes, we can argue that compromise is at the heart of an inclusive politics, but to imagine that there is a halfway point that we can define as the 'sensible centre' is entirely the wrong metaphor, and to rely on it as an organising principle – as a *political* principle – is to be led astray. I can illustrate the point with a simple example we are all familiar with: same-sex marriage.

In the recent past, the issue was pretty much settled, or so it seemed. A majority saw marriage as something that happened between a man and a woman, and the very notion of same-sex unions was considered extreme. The sensible centre was: no marriage for you, same-sex couples. However, we now know that the 'centre' on this issue has shifted, that a majority of people in Western democracies now support same-sex marriage. The reason opinion has changed is because the people who were previously described as extreme in their views continued to argue their case and were able to change the conversation: they made it less about people of the same sex marrying each other and more about two people loving each other, regardless of their sex. This *new* sensible centre on same-sex marriage was not some uncontested, pre-existing condition, not some halfway point of compromise waiting to be reached, but something that had to be constructed over many years.

Yes, we can decry the partisan nature of some of that argument, especially its vitriol and violence, but in a democracy we don't get to bypass it altogether. For heaven's sake, that *is* democracy!

This is an important framework for the ideas I will present in this book. Progress does not revert to a mean. It is infinitely contested and therefore has to be fought for. To aim merely at 'the centre' is a rookie mistake, like aiming at a bullseye in darts when the high score is actually the triple 20 at the top edge of the board. Recent work by US political scientist David Broockman found that even when people identify as 'moderates', they actually tend to hold at least some views at the so-called extremes. What happens is that such extremes are averaged out in polling data and thus the 'moderate' or centre position is nothing much more than a statistical remnant. The so-called middle is always – always – a contested political position, and to call it 'sensible', as if all wisdom necessarily resides there, is to engage in a sleight of hand.

It is reasonable to wish for a less cantankerous debate, one with less abuse and less vitriol. It is reasonable to wish for fact-based discussion, where expert opinion is used as a guide to democratic decision-making. It is also reasonable to wish for a media that better reflects the variety of views on any given topic, and one that is less beholden to shallow caricature. But it is completely unreasonable to define yourself as a moderate and *your* position on everything as the sensible centre and to presume that nearly everyone else is therefore extreme.

This brings me to the heart of this book.

What I set out is a series of proposals that will reorient our societies around the principle of a life in common. Not a centre, but a life in common. And by that I mean, basically, more public ownership. Public space. Public activity. Joint projects and the greater good. Equality. Fairness. A fair go. Access to essential services independent of market price. Bottom-up change. I am not against private property, of owning things ourselves as individuals, and I don't believe in some Borg-like collective. I am not advocating for full-scale central planning and exact equality of outcomes. But a democratic society can only function properly if its citizens share in everything from power to healthcare to government itself. It's how we manage risk and thus make life viable in complex societies. Remove these things from common ownership and by definition you hand control from the many to the few. Once you do that, you undermine the very notion of democratic politics because the basic question of democratic politics – who gets what – is no longer a collective decision but a market decision. It is decided by those who control the money rather than by citizens.

This life in common is important for another reason: our individuality is formed through collective existence, not the other way around. In his *Theory of Moral Sentiments*, Adam Smith, the founder of modern economics, recognises this fact:

> Were it possible that a human creature could grow up to manhood in some solitary place, without any communication with his own species, he could no more think of his own character … than of the beauty or deformity of his own face … [H]e is provided with no mirror which can present them to his view. Bring him into society, and he is immediately provided with the mirror which he wanted before.

Our collective existence relies on spaces and services owned in common, things belonging to no-one and to everyone, the benefits of which accrue to us all equally by virtue of our membership of a society. Yet we have spent the last 40 years selling off nearly everything we own in common, from public parks to education services, to power grids and public transport, and now our governments are filling up with – or doing the bidding of – oligarchs and authoritarians. Big surprise!

The net result has been a loss of control by ordinary people over their own lives, a collapse in trust of our political parties and government institutions, and a situation where our much-vaunted individual freedom has turned into nothing more than a hierarchy of wealth and power with many of us left behind. Inevitably, there has been a parallel rise of a divisive politics. In the absence of a life in common, of shared control over the way our societies develop, people retreat to tribalism, and that is when certain politicians come along to exploit this and create bogeymen for us to fear, from asylum seekers, to Muslims, to the unemployed. Without a healthy commons, everyone is potentially the enemy of me and mine.

So that's where we have to start: repopulating our

commons. Reinventing and – if necessary – buying back our common sovereignty. Recognising that we must be individuals but that, paradoxically, we cannot achieve that individuality alone.

The first section of the book, Premise, sets out the underlying ideas and presumptions about why we need to make these societal changes and, just as importantly, how to make them happen. I am sick of reading books or articles that put forward wonderful, challenging ideas for change and simply assume that their good intentions are so obvious a majority of people will automatically go along with them. Make no mistake, I am offering some pretty challenging ideas myself. But I offer them in the firm belief that they will have to be fought for.

Since no theory of change is meaningful without a theory of power, I begin with a chapter about power and the role of politics, before moving on to discuss in more detail the idea of 'the commons' that underpins my entire argument.

We cannot expect to build a decent society – one that is peaceful, fair, inclusive, and which provides opportunities for everyone to thrive – unless we build institutions that reflect those values. The success of the nations that have embodied this type of society since the Second World War has been built on such institutions; we were able to build them because there was broad commitment to underlying values such as individual freedom at the personal and economic level, a distribution of wealth that tended to equality, an education system that not only prepared people for work but that inculcated these values, and most of all, systems of government that as much as possible represented those who were governed. Having emerged from a period of war and instability, the need for such institutions was almost self-evident. In his book *On Tyranny*, Timothy Snyder says: 'It is institutions that help us to preserve decency [but they] fall one after the other unless each is defended from the beginning. So choose an institution you care about – a court, a newspaper, a law, a labor union – and take its side.' That is what I do in the second section of this book, Practice: I choose five institutions (broadly speaking) that I not only defend, but for which I suggest improvements, ways we can rebuild, reinvigorate and rethink them from the ground up so that they better fulfil their basic functions. They are media, government, work, wealth and education. In choosing these five, I am not saying that they are the only things that matter or need our attention, and indeed, you may well place your priorities elsewhere. My reason for choosing them is simply that they are the ones I have thought about the most and can say something useful about. Besides, this is meant to be the start of a conversation, not the end.

After moving you through each institution, the final chapter is on joy, where I try and give a sense of the deep sense of happiness that is possible when we build a life in common.

So the book begins with a theory of power, and ends with a theory of joy.

Politics is ultimately about one fundamental problem: how do we balance the needs of the individual against the necessity of living together? It doesn't really matter whether 'living together' refers to a city the size of Beijing or a commune with six families living in tents. At the end of the day, all the problems come down to individual versus collective rights. Sure, the bigger the settlement, the more people involved, the more complex the problems are likely to be:

Beijing is harder to run than six tents. But at its heart, the issue is always how do I, as an individual, with all my quirky needs and wants, manage to live together with a bunch of other individuals, who have their own equal and opposite quirky needs and wants, without us all killing each other?

Every major religion, belief system or political theory is an attempt to walk this tightrope. So is this book.

When you stop worrying about the future, you fuck up the present. Those who deny climate change, or who insist that the earth is God's gift to us to do with as we wish, or who assert that corporations are people and that a business's sole aim should be maximising shareholder wealth in the short term, have at some place deep in their psyches decided that the future doesn't matter, that either we all die and go to heaven or that we all just die. And so their only goal is to rig things in their own favour for the term of their natural life. Right now, those sorts of people are in ascendancy and they are going to be hard to overcome.

The ideas offered here are designed to be achievable within the current political framework without having to tip over into revolution. None of them is easy to do, let alone inevitable, but they fall within the parameters of the possible. Unless we can come together and do these hard things, or things like them, we are going to lose much of what we take for granted and hold dear. If we just trundle along and think that business as usual will be enough, we are in for a rude awakening.

George Orwell, in his 1941 essay 'The Lion and the Unicorn', wrote a sentence that blew my mind when I first read it (and still does): 'As I write, highly civilized human beings are flying overhead, trying to kill me.' He continued:

> They do not feel any enmity against me as an individual, nor I against them. They are 'only doing their duty', as the saying goes. Most of them, I have no doubt, are kind-hearted law-abiding men who would never dream of committing murder in private life. On the other hand, if one of them succeeds in blowing me to pieces with a well-placed bomb, he will never sleep any the worse for it. He is serving his country, which has the power to absolve him from evil.

Let me update Orwell's formulation.

As *I* write, highly civilised human beings are sitting in comfortable offices figuring out ways to deprive people of the means for a decent life. They are 'only doing their duty', as the saying goes. Most of them, I have no doubt, are kind-hearted law-abiding men and women who would never dream of robbing someone in private life. On the other hand, if they succeed in halting a welfare payment, or saving the government money by closing down a public hospital ward, or depriving an unemployed person of a week's dole payment because that person didn't tick the right box, they will never sleep any the worse for it. They are serving their government, which has the power to absolve them from evil.

It is in everyone's interests, *especially* those with the most – the most power, the most money, the most influence – to recognise their real self-interest, that of ultimate survival. We are pushing hard against the limits of what we can extract from the earth and it is causing all sorts of problems, political and social, economic and environmental, everything from global warming, to mass extinctions, to mass migration. We have become brilliant at many technical things, but we are really bad at recognising what is going on in front of our eyes, stepping back and doing something sensible about it. This stuff is not rocket science. It is politics, so much harder. But if we can get the politics right – via the sorts of measures outlined in this book – then something great is within our grasp.

A driving idea behind this book is that the future is unknowable, that we need to be humble in the face of that, and however well thought out our schemes for improvement are, they need to be open to constant adjustment. The idea in producing this particular set of suggestions is not a demand that they be adopted, and certainly not a suggestion that they represent the one true way. They are offered merely to add to the ongoing and infinite discussion that is democracy, to be picked over and improved by my fellow citizens.

British author Raymond Williams once claimed: 'To be truly radical is to make hope possible rather than despair convincing.' That's all this book is trying to do.

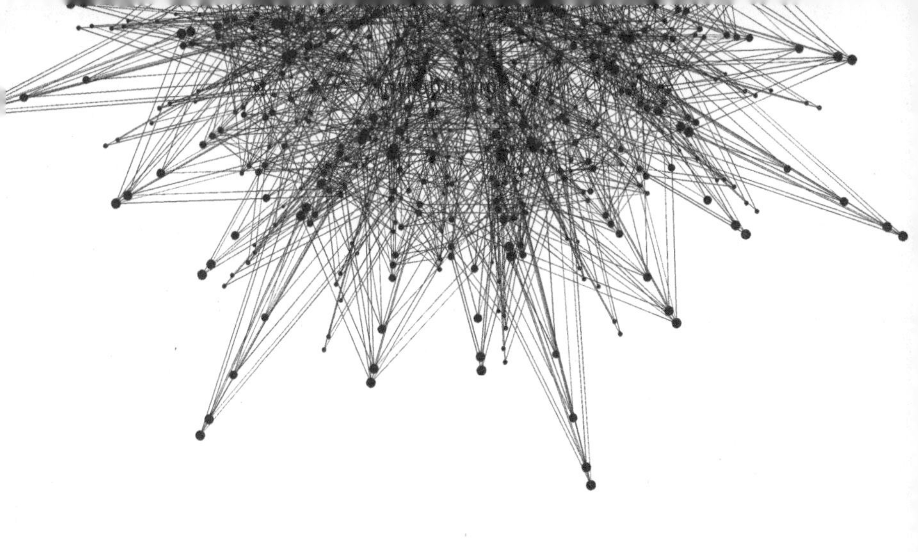

PREMISE

POWER

> If the smartest guy in the room has no connection to power he's just an annoying smart-arse.
>
> *Former ACTU assistant secretary Tim Lyons*

We have become a 99–1 society where the richest few control the vast majority of the wealth and resources of the world, and where they wield in their own interests the political power that comes with that control. When, as Oxfam reported in 2016, 62 individuals own as much wealth as the poorest 3.6 billion people in the world, something has gone wrong. Democracy has the tools to fix this, but only if we are willing to use them, and I think that is what's missing from current debates. Progressives – people who believe in a fair society and a life in common – have become afraid of power. We are more interested in winning arguments than kicking arses, and that has to change. If you want to beat the 1 per cent, smart ideas and clever analysis are not enough: you have to wield power.

When inequality becomes entrenched like this, and people can see no other way out, real violence threatens. In his book *The Great Leveller*, Walter Scheidel points out that across millennia only four things have ever led to a more equal society: 'mass mobilization warfare, transformative revolution, state failure, and lethal pandemics'. Only when one of these catastrophes destroys the current order, Scheidel says, do we have a hope of a more equal world. He makes a compelling case, but it shouldn't stop us from trying to break the pattern.

You want to avoid violence? Then you'd better learn how to wield power. You want to stop feeling depressed and overwhelmed and hopeless in a world threatened by climate change, growing inequality and a political system that seems increasingly useless? Learn how to wield power. That means getting together with others and making things happen. It means working methodically and gradually towards ultimate goals with those already in power, and it means resisting their resistance. When they thwart one action, it means being willing to escalate, to keep going, to get more creative. Ultimately, wielding power is about grassroots organisation – and we will get to that – but first we need to recognise who currently wields power and how they do it.

We all understand who the powerful people are in our society, though the nature of their power isn't always clear. Politicians, bureaucrats, the police, our bosses all have, in different ways, some power over us and can cause us – directly and indirectly – to do what they want. We understand that we have to obey the law because the law enforcers may punish us if we don't. We mightn't think of it in this way, but we know they are able to punish us because they have a monopoly on the use of violence. In political science, this monopoly over violence is the defining characteristic of the state. German sociologist Max Weber was probably the first to define the state in this way. In his essay 'Politics as a vocation', he writes: 'If no social institutions existed which knew the use of violence, then the concept of "state" would be eliminated, and a condition would emerge that could be designated as "anarchy".' Therefore, he says, 'a state is a human community that (successfully) claims the monopoly of the legitimate use of physical force within a given territory'.

His use of the word 'legitimate' is key. In a democracy violence is constrained, but we know that if we resist the power of the state, or of certain institutions within it, that violence will be used against us. We know, for instance, that state violence threatens us when we take to the streets and protest in support of ideas or actions, go on strike, or resist any other sort of injustice we perceive. We know that if we show up on the shores of another country without the proper permissions, violence is likely to be waiting for us, as the millions in refugee camps and detention centres can attest. To the extent that we are happy for the law to be enforced in these situations, then the state's monopoly of violence is considered legitimate. But democratic institutions are based on the idea of checks and balances – institutional controls like the separation of powers and the ability of one house of parliament to review the laws enacted by the other – so that we, as citizens, are able to monitor how the state wields power to make sure it stays that way. To quote political theorist Hannah Arendt, power doesn't need justification, it simply is 'inherent in the very existence of political communities'. What it does need is legitimacy.

Not all the power exercised in a society involves the threat of violence; nor does it reside entirely with the state. So while overt power is relatively easy to understand, our conception of power needs to be more nuanced. We know, for instance, that wealth buys influence and that the rich have the ears of politicians in a way that ordinary citizens don't. Corporations employ lobbyists to convince politicians to enact laws that will favour those corporations. (Non-government organisations such as charities or even sporting groups play a similar role, though their aims are generally more altruistic.) Other organisations such as think tanks exist to help shape public discussion about major issues, everything from education to social security to defence to government itself. There are right-wing think tanks and more progressive ones, and via their research and their reports they try to exert power over the way in which society is understood and run. To the extent that these organisations are funded by particular sources, the reports they write are likely to be influenced by the views of those donors. Such uses of power put a strain on its legitimacy, because they can be beyond our ability to monitor and consent to.

In Australia one right-wing think tank in particular is closely associated with the Liberal–National Party Coalition, which has formed government more often than any other party since the Second World War. That think tank is the Institute of Public Affairs, or IPA, and it was instrumental in the formation of Liberal Party itself. The IPA has traditionally seen its job as helping muster opposition to organised labour, as opposing the Labor Party itself, and it has traditionally been funded by various corporations and other representatives of capital.

The Labor Party, on the other hand, is an offshoot of the organised labour movement that pre-dates it. The union movement has been a powerful force in our society, though its power has waned over the last four decades. Nonetheless, the contest for political power in Australia has largely been one between the conservative and the Labor parties. Each in turn is influenced via the veins of power that flow from the organisations of civil society, like the IPA and the union movement, and also from smaller organisations and even individuals, trying alone or together to exert some power over what sort of society we have. This kind of power ecosystem is common throughout the nations of the Western world.

Increasingly, the two-party political system that operates in most developed democracies is breaking down, and we have seen the rise of independents and smaller parties, many of which reflect the views of particular segments of the population. Using Australia as an example again, we can point to the success of political candidates who represent groups such as shooters and automobile organisations. Independents like Nick Xenophon and Pauline Hanson are able to rally sizeable support around single issues such as anti-gambling and anti-immigration, even if their agendas then broaden. Most successful of all have been candidates representing the Greens, who tap into community support around environmental issues and then leverage their success to advocate for progressive policies across a range of areas. The electoral success of most of these sorts of candidates is relatively volatile, suggesting the sources of their power are also volatile and therefore limited. Nonetheless, their very existence points to the possibility for those outside of elite groups being able to exert some power, and we should take hope from that.

Other key sources of power in Western democracies are the media and the Christian church. The media are important and complex enough to warrant their own chapter, and that is coming up. While the church's influence is waning in most Western democracies – less so in the United States – legal exemptions for religious institutions give them freedom from the constraints imposed on other large organisations such as trade unions. The Catholic Church in Australia holds $80 billion in assets, has its own bank and superannuation fund, and employs 220 000 people; but it pays no tax and cannot be sued, and its internal operations are not open to public scrutiny. That is power. But of course, other religions are not imbued with the same power within the Western state. Islam is systematically and structurally demonised – at the very least marginalised – and this marginalisation extends to other minority groups too.

This raises the hidden power of *advantage*, the sort of power that exists simply because you are the member of a particular group or faction, or that leaves you vulnerable because you are not. Across the fields of religion, gender, race, physical ability, sexual orientation and class, power operates differentially. Advantage – *power* – inheres in these categories and they intersect: you can be doubly or triply advantaged or disadvantaged, as the clusters of 'rich white man' or 'poor black woman' amply suggest. Class intersects with all of these, and there is certainly value in recognising that not all white men are powerful and that not all black women are not. But there is an ongoing and misguided argument that then arises as to which identity we should prioritise.

There are those on the left, and among progressives more generally, who see class as the ultimate unifier, that no matter what other categorical or structural disadvantages apply, it is our battle for economic equality that should be front and centre: the poor black woman, they say, has more in common with the poor white man than she does with the rich white woman. But I am not convinced it's that simple. Class is clearly an important marker of how much power a person has in our society, and it needs to be central to the discussion, but in a world where work itself is changing and the collective identity of a 'working class' is no longer as reliable a way of categorising people as it once was, we have to be careful not to be reductive. Attempts to conjure a new working-class category like the 'precariat' – those in the modern labour market who are increasingly called upon to do insecure, part-time, casual or self-employed forms of work – are worthy, but they do not cancel out the other identifying categories of race, gender, religion. They always intersect. Author and commentator Alison Croggon, writing in *Overland* magazine ('On power') says, and I agree:

> a special pox on those who use the term 'identity politics' as a catch-all to marginalise various movements that are attempting to address the structural bigotries of colonial, capitalist, patriarchal society. The struggles against racism, sexism, homophobia, transphobia and ableism are all demands for justice that stem from searing suffering, and they all have common aims. But too often their particular struggles are dismissed as of secondary importance.

One of the arguments against identity politics is that it invites the enemies of progressive politics to summon *their* identity and claim special favour for themselves. Writing in *Jacobin*, Shuja Haider ('Safety pins and swastikas') argues for a 'politics based on a universal principle, rather than on the opposition of identities'. He suggests that although 'left-liberal identity politics and alt-right white nationalism are not comparable' they are nonetheless 'compatible'. The 'adoption of the language of identity politics' by the right, Haider says, allows them 'to draw the battle lines, marking the territory of their white national fantasy'. But surely the right's cooption of identity politics merely underlines that group's inherent advantage, their power to do so, and so to shy away from 'identity politics' capitulates to that inherent advantage. Under such circumstances, *why* emphasise this alleged compatibility? Why not stress their incomparability?

The identity politics of #metoo or #blacklivesmatter or #equalmarriage are a call for inclusion, for equal treatment and freedom from prejudice. By calling attention to abuse and inequality, they are not intended to divide but to unite. They are the elements of which universality is ultimately made: they contribute to it; they don't contradict it. They are about demanding that our institutions accord them the same treatment that more privileged groups take for granted. This is the exact opposite of asking for special favour. So when misogynists and white supremacists run around claiming victim-status for themselves, asserting their identity and their 'right' to have that identity respected, they are not seeking equality but instead a return to, or continuation of, the privilege they already enjoy. This sort of identity politics is strategic and phoney, designed to prescribe for themselves special rights and exclusions and to actively oppress anyone who is different from them. It is a rearguard action against their own loss of power. The claims by the alt-right and their myriad brothers-(and they are mainly brothers)-in-arms may be centred in identity, but they are aimed at completely different ends to those of left-liberal groups.

The repulsive nature of this sort of right-wing, white supremacist, male-dominated form of identity politics doesn't undermine the legitimate claims of people of colour, the LGBTQIA community or the religiously marginalised; in fact, it simply highlights why the claims of those latter groups need to be at the centre of any progressive movement that seeks to neutralise the hidden power of advantage that pervades all our institutions. Class is not enough, because people's experience of oppression is felt in other aspects of their identity, and to ignore that is to validate the right's attempt to hijack the matter of identity. As Croggon says, 'if a collective movement reproduces the prejudices of the society it claims to be trying to improve, it inevitably sows the seeds of its own disintegration'.

Most discussions of power focus on how it resides among society's elites, but it is important to realise that in democracies, even ones as damaged as most currently are, a logic of power exists that even now allows for elites to be challenged. The norms of democratic governance – that we are all equal before the law, that we all get a say, that a certain level of fairness is expected – may be violated and violated often, but their existence is itself a kind of power in that it shapes expectations of behaviour. (Part of what makes Donald Trump unique among political leaders of recent years, and potentially so dangerous to democratic society, is his ability to flout the norms of political behaviour with, at this stage at least, impunity.) Power isn't just the ability of elites to make others do their will but also exists in the fabric of our institutions and our relationships. It shapes the world we live in, often without us being conscious of the process.

I am laying all this out to reinforce how stupid it is to think that we can enact any sort of progressive agenda without an understanding of how power works and a willingness to use it. Nice ideas are not enough. They have to be encoded in our institutions so that they become part of the background operating system, so to speak. Once that happens, they are powerful almost by definition, because they form the day-to-day common sense of the workers in those institutions. They no longer have to be explained and argued for: they are just there. This means you need the ideas in the first place, but they are useless without a theory of power that allows you to install them ahead of other ideas. In fact, institutions are often built – or undermined – with an idea, or an array of ideas, and so progressives need to be powerful at that level too. But I will discuss that in the next chapter. In this one, I simply want to stress that good ideas are a necessary but not a sufficient basis for achieving political ends.

All of this means not only that we have to work within existing structures and institutions – we do – but that we have to be willing to create new spaces where ordinary people can learn the habits of power and the confidence to use it. Ironically, progressives are looking to the success of the right wing itself for models of how to do that. Many of the major analyses of neoliberalism, for instance, document how these ideas insinuated themselves gradually into the body politic. Books such as *Dark Money* by Jane Mayer are a deep look at the way people like the Koch brothers use networks of think tanks and 'charities' to not just influence the legislative agenda in their own favour, but to build grassroots support for their ideas, even as those ideas work against the interests of most people.

In *Inventing the Future*, Nick Srnicek and Alex Williams lay out the process by which neoliberalism captured state elites via the backchannels of think tanks and academic conferences. Beginning with the Mont Pelerin Society (MPS), founded in 1947 and led by economist Friedrich von Hayek, those associated with the movement began the long, slow game of 'changing political common sense' in order to 'develop a liberal utopia' by establishing think tanks propounding neoliberal ideas and placing the movement's supporters in government positions. Srnicek and Williams point out that these early neoliberals consciously avoided 'folk politics by working with a global horizon, by working abstractly (outside the parameters of existing possibilities) and by formulating a clear strategic conception of the terrain to be occupied', which they saw as a way of influencing elites, not the general public. Although 'capitalists did not initially see neoliberalism as being in their interests', the MPS was ultimately pushing an open door, and so in the time between its founding in 1947 and the economic crises of the 1970s – which brought into doubt the principles of Keynesianism and the economic order based on them – the ideas of neoliberalism were clearly installed, at least among these elites, as a ready-made alternative.

And this is worth bearing in mind, too: crisis can be a powerful force for change, showing the shortcomings of established ideas in a way that no amount of logical argument ever can. But this doesn't mean that crisis inevitably leads to progressive outcomes, far from it. In fact, elites will often generate their own crises to help them implement policies and programs that they see as beneficial. In her book *No Is Not Enough*, journalist Naomi Klein calls this approach the 'shock doctrine', which she describes as 'systematically using the public's disorientation following a collective shock – wars, coups, terrorist attacks, market crashes, or natural disasters – to push through radical pro-corporate measures, often called "shock therapy"'. Klein argues that President Trump's attacks on the regulations that keep corporate greed at bay 'can be counted on to generate wave after wave of crises and shocks'. But crises or shocks alone are not enough. Klein adds:

> A state of shock is what results when a gap opens up between events and our initial ability to explain them. When we find ourselves in that position, without a story, without our moorings, a great many people become vulnerable to authority figures telling us to fear one another and relinquish our rights for the greater good.

In recent times a number of right-wing groups have been successful in influencing the elite institutions of the state, and progressive groups such as Indivisible in the United States are now consciously using right-wing models of activism as a guide. Indivisible sets out on its website the way in which the Tea Party, for instance, came to dominate the Republican Party, noting that success came from two 'critical strategic elements'. First, they were locally focussed, often meeting in groups of fewer than ten people to coordinate advocacy efforts. They consisted of a 'relatively small number of groups [that] were having a big impact on the national debate'. Second, the Tea Party was 'only defensive', focussed on criticising the Obama administration's policy rather than proposing new policy. They specifically targeted 'weak Republicans', those likely to make concessions or otherwise work with Obama, attending local political meetings to demand answers and organising mass phone-ins to congressional members when key votes were pending.

Author Thomas Frank is another who holds up the Tea Party as a model for progressives to understand. In his assessment of what went wrong with the Occupy movement (the closest US progressives have come to forming their own Tea Party), Frank is angry at what he sees as the failure of Occupy to move beyond lofty ideas and convert the momentum they had into political power. In his *Baffler* article ('To the precinct station') he chastises the theoretical approach the Occupy protesters took, saying it does not 'require poststructuralism-leading-through-anarchism to understand how to reverse these developments. You do it by rebuilding a powerful and competent regulatory state. You do it by rebuilding the labor movement. *You do it with bureaucracy.*' The Tea Party might've started with the same claims of having a horizontal structure and no demands, but unlike Occupy, this was a strategic ploy not an inviolable principle. Instead, he says, 'the Tea Party didn't really mean any of its horizontalist talk ... [it] had no poststructuralist thinkers contributing to theory magazines.' What it did have was

> money, organization, and a TV network at its back. It quickly developed leaders, and demands, and an alignment with a political party ... And perhaps that was the plan of the movement's masters all along. The vagueness and the leaderlessness were merely for show ... gimmicks designed to give the product the widest possible appeal in the early days.

From the Mont Pelerin Society to the Tea Party movement, one lesson to be learned is that a small but committed band of people can bring power to bear on an entire elite system. If the right actions are generated at a grassroots level, or directed at the right set of elites, a relatively small number of people can make a difference. Fascinatingly, the same point is made by Roger Ailes, the former Nixon adviser hired by Rupert Murdoch to run Fox News, and who turned that organisation into probably the greatest propaganda machine since Goebbels. In 2017, in an interview with reporter Michael Wolff for *The Hollywood Reporter*, Ailes says the success of Fox News was predicated precisely on the fact that it appealed to a small number of devotees rather than a mass audience. (Of course, this analysis has to be taken in the context that Fox News was set up with a media billionaire's resources, but the point still stands). What makes it doubly interesting is that he says the same principle applied to the success of Donald Trump, telling Wolff that Trump's 'election proved the power of a significantly smaller, but more dedicated electoral base – as, likewise in cable terms, a smaller hardcore base was more valuable than a bigger, less-committed one'.

US organiser Jane F McAlevey makes the same point in her book *No Shortcuts*, which analyses some successes of the US labour movement in recent years. She says that in many ways the 'fight to save education and health care [or whatever else] is a fight against the logic of neoliberalism'. She is critical of the willingness of the union movement (to which she belongs) to seek gains for workers by entering into accords with businesses, rather than engaging in more grassroots-based campaigns. She argues that while these agreements buy some short-term benefits, they ultimately enter into the logic of neoliberalism and therefore work against the long-term interests of workers. 'Corporate collaboration isn't new', she says, 'but when the labor-run corporate campaigns first developed in the 1970s as a response to the degeneration of worker protections under US labor law, they were designed to complement worker organizing. By the early years of the new millennium, they had all but replaced it.'

The Hawke–Keating government's accords, beginning in the 1980s in Australia, are a good example of this, and it is no coincidence that the same government introduced the reforms that opened the neoliberal floodgates in Australia. Although Hawke and Keating predicated their accords and other reforms on increases in the social wage – a near-universal healthcare scheme in the form of Medicare, a superannuation scheme, and other gains for workers and for society more generally – these safeguards have not been enough. The logic of neoliberalism, comprising competition, individual entrepreneurialism and private gain over public wealth, has been overwhelming and has crowded out the benefits of the social wage Hawke and Keating put in place.

The problem, McAlevey argues, is that the labour movement lost sight of the idea that workers – or any other relatively powerless group in society – need to be involved in their own emancipation, and that 'corporate campaigns' for workers' rights have instead 'resulted in a war waged between labor professionals and business elites'. Central to any campaign that seeks to exercise power on behalf of we-the-people, McAlevey says, is 'a careful, methodical, systematic, detailed analysis of power structures among the ordinary people who are or could be brought into the fight'.

Ultimately, your aims will be achieved only if these people are leading the process of change. 'Large numbers of people transition from unthinking "masses" or "the grass-roots" or "the workers" to serious and highly invested actors exercising agency', she says, 'when they come to [understand] and to value the power of their own salient knowledge and networks.'

Many progressive campaigns fail because they are detached from those they most affect, run instead by highly educated leaders who themselves are comfortable in the halls of power they are trying to challenge. A number of problems arise because of this. One is that engaged progressives tend to see themselves as much better informed politically than those on whose behalf they fight, especially around matters of race and gender, and therefore fall into the habit of chastising anyone who makes a comment they see as racist or sexist. Writing in *Overland* ('We had Marx, they had Pauline'), Joanna Horton says that 'while "calling out" someone's racism feels good (or at least righteous) in the moment, it's highly unlikely to change their mind'. She speaks of her experience campaigning with the Anti-Poverty Network (APN) in regional Queensland and how activists try to find common ground. She quotes an organiser who says, 'You need to create something people want to buy into and people don't want to buy into being made to feel like dirt. They're already made to feel like dirt because they're poor.' The tendency to chastise rather than empathise chases people into the arms of right-wing populists like Pauline Hanson. The APN organiser says,

> if you really listen to the reasons why they're going for Pauline Hanson, you see that they've come to the

> same realisation we did. It's just that they didn't have the same solution in front of them. We had Marx, they had Pauline. They'll tell you, I don't agree with everything she says. But she's real.

Horton concludes by saying 'political organising doesn't have to be ... based on attitudes'; in fact, it shouldn't be. She says, instead, that it should be informed by 'the material conditions of people's lives – the things they're already experts on, things they don't need to be persuaded to care about'. At the end of the day you have to 'build an organising model based on mutual interests, on building solidarity, and on politicising poverty without fear'.

Another issue that arises because of this disconnect between leadership and grassroots is that there are endless demands from elites and technocrats for things like 'evidence-based policy' or 'better data' or for 'reframing the debate'. These methods may have some value, but all are built on the idea that ordinary people are there to be led or convinced, or even bullied and condescended to, rather than made integral to the process itself. The idea of 'reframing' comes up a lot and is a particular bugbear of mine. It is largely associated with US academic linguist George Lakoff and his book *Don't Think of an Elephant!* There is nothing wrong with his overall argument that language encourages us to think of things in particular ways, and so if you can change the language – reframe the debate – you can change the way people think about the subject at hand. He says, probably correctly, that the right wing is particularly good at using frames such as 'death taxes', 'war on terror' and 'family values' to steer the argument in certain directions, while progressives are bad at it and need to improve.

This is a useful insight – *of course* it matters how we talk about things, the words we use – but such an approach tends to appeal to those who like to analyse problems rather than solve them, and what's worse, it puts the cart before the horse. The power of words, or frames, should be part of our analysis, but simply telling people to come up with new frames is not a strategy of power, it's a parlour game. What Lakoff and others who push the idea of reframing tend not to notice is that the reason right-wing frames work, or become the accepted form of words, is because they are embedded in systems of power that already lean that way. So it is those underlying power structures you have to change, not just the language they use. Indeed, successful frames *emerge* from these underlying structures. It isn't as if someone came up with the phrase 'war on terror' or 'death taxes' and they, by themselves, created the powerful associations that work in conservatives' favour. The associations *already existed* and the phrases themselves were organic to the process, not imposed from outside. When conservatives tried to reframe the privatisation of social security in the United States as the introduction of 'personal accounts', however, it failed miserably because most voters actually prefer government-provided social security. Reframing only works when momentum – power – is already with you. It follows that progressives cannot simply substitute alternative phrases and expect them to gain momentum without first changing the underlying power structures.

Data is another blind alley. That is, the idea that if we do enough polling and collect enough demographic information about voters, we will unlock the secret of influencing them. As with framing, I won't dismiss it altogether: it is a vital tool. But I do want to tip a bucket on the idea that it will solve all our problems. The reason is that data alone – like framing – sidesteps engagement with the people who are meant to benefit from progressive policies. It turns them into a subject to be manipulated rather than a resource to be developed and deployed on their own behalf. So, as useful as it is to build databases and create profiles of citizens and, perhaps, use those to target particular messages at particular audiences, that can't be a substitute for the direct involvement of those a given policy is designed to benefit. Even where data crunching has been successful, as in the Victorian state election in 2014 that brought the Labor government to power, it was not used as an end in itself. There, big data was used to identify sympathetic people within particular electorates and encourage them to carry the message to their neighbours. ALP assistant state secretary Stephen Donnelly, who ran this aspect of the campaign, told *The Australian* newspaper that data was used, but the key was to build 'a network of more than 5500 volunteers and 250 volunteer leaders' who would then go out into the community and talk with people. Donnelly said that the secret lies in 'the devolution of power – a focus on values rather than party connections. Many voters have stopped listening to political leaders ... and the most powerful messages are those delivered by everyday people who echo their values.'

Again, becoming powerful and thus being able to change things comes from working closely with those most likely to be affected by the proposed changes. McAlevey suggests three basic approaches can be taken: advocacy, mobilising and organising. But while the first two have their uses, it is only organising, she argues, that leads to deep and substantial change, and this is the key to progressives understanding how they might exercise power in an environment like the modern democratic nation. She is blunt about it: 'In unions and [social movements] ... advocacy and, especially, mobilizing prevail. This is the main reason why modern movements have not replicated the kinds of gains achieved by the earlier labor and civil rights movements.'

An important difference between organising, advocacy and mobilising is the amount of pain each can inflict on those who resist change, which is ultimately the key to success. In the second section of this book I talk about some specific reforms, including shorter working hours and employee ownership of firms, a universal basic income and a parliament in which the members are selected through a lottery process. All of these are major changes to how we run society and, if implemented, would change the way power is distributed in the various institutions involved. They would then assume a power in their own right, become part of the background operating system, and influence the sorts of decisions made in the future. My point here, though, is that none of this can happen without first imposing some costs on those who run things now. McAlevey draws on the work of civil rights leader Joseph Luders, who wrote the book *The Civil Rights Movement and the Logic of Social Change*, in which he argues that even when dealing with the racists who fought against civil rights, success most often came when those racists met with some sort of economic hardship because of their resistance to change. Basically, there are two types of costs that can be imposed, which McAlevey calls the 'cost of settlement' and the 'ability to create a crisis'. She says that although the two costs are always related to each other, the power to create a crisis is how we measure what needs to be done to bring about change. It is therefore important to do a full 'power required' assessment in any given situation. Shying away from such assessments is what has made progressives weak, has made reform difficult and has left the field open for the 1 per cent, who have absolutely no qualms about imposing their own costs on the rest of us by creating a crisis whenever necessary.

It is when we baulk at this sort of *organising* that progressives tend instead to embrace advocacy and mobilising.

Another difference between advocacy, mobilising and organising is the way in which each involves ordinary people in the process. McAlevey says that advocacy doesn't use ordinary people at all and prefers 'lawyers, pollsters, researchers, and communications firms' who are employed 'to wage the battle'. This comes from a tacit belief among such advocates that elites will always rule and that the best you can hope for is to install 'better' elites, those more aligned with your values. It is similar to those who insist on technocratic solutions to political problems, the idea that if you can just bring enough expertise to bear, everything will be okay. The trouble is that technocrats focus on means rather than ends, and often the ends are what's at stake. McAlevey says, the '1 percent have a vast armory of material resources and political special forces, but the 99 percent have an army', and the problem with advocacy – the preferred tool of well-intentioned people of the centre left – is that it leaves that army at home. Advocacy leads to closed-door deals between elites – like accords – that tend to be imposed on those they affect. Such accords may be broadly accepted but excluding the base from the negotiating process leaves such settlements vulnerable to incremental change and ultimately disbandment.

Mobilising is better in that it seeks to involve large numbers of people – by rallying them for mass demonstrations and the like – but, McAlevey says, 'too often they are the same people: dedicated activists who show up over and over at every meeting and rally for all good causes, but without the full mass of their coworkers or community behind them'. Such people see themselves as the 'agents of change' and are therefore more interested in rallying their own than including others. In this, social media and other online tools can be useful, but again, they can't be made to substitute for actual organising. I am deeply sceptical of claims that Jeremy Heimans and Henry Timms make in their book *New Power*, lauding the way in which the internet breaks down barriers to the various forms of knowledge and information, thus creating 'new power'. While this is indeed a huge advantage of online networks, it's not enough. Such networks may allow certain groups and individuals to influence decision-makers in ways that weren't previously possible, but they are far less likely to enable real structural change. And even where that might be possible, there is nothing intrinsic about online activity that prepares a new cohort of leaders to emerge.

Organising, on the other hand, is *predicated* on the involvement of ordinary people. They are not just there to make up the numbers. The whole point of organising is to identify and train leaders from the base so that they can take over and run the operation. Organising, in McAlevey's words, 'places the agency for success with a continually expanding base of ordinary people, a mass of people never previously involved, who don't consider themselves activists at all'. Identifying leaders within communities is more effective than training volunteers because these 'organic leaders' bring with them ready-made numbers. This approach is also built on recognising people not just as one-dimensional members of a given group – say, workers in an industry – but as multidimensional members of a community, complete people whose needs and wants are diverse and complex and must be satisfied on many levels. Advocacy and mobilising can become a case of preaching to the converted, with the same people showing up to meetings and rallies or following each other on social media: *organising* forces movements into the broader community, reaching those who might not normally involve themselves in actions designed to create change. The success in Australia of the equal marriage plebiscite was built on activists reaching out to ordinary families and community members over many years. These actions brought people into contact with LGBTQIA people in a way that might not have normally happened, which allowed them to discover the things they had in common rather than focussing on difference. Eventually the plebiscite that decided the matter was the most successful ever held in Australia: 133 of 150 electorates voted yes and every state in the nation returned a yes majority.

Many in Western society right now share an enormous mood for change and an understanding that we can't go on as we are. What's less clear is how to bring that change about. Organising – wielding power from below – is the key to the change progressives want, and they need to embrace it. Organising creates solidarity that endures beyond the immediate campaign and so helps cement the gains made. What's more, it recognises that what is often at stake is ends and not means. Having command of the means – the expertise – doesn't give you a right to determine ends. In other words, while it is important to get the technicalities of implementation right and to recognise that experts have a role to play, it is ordinary people who should have the final say about what the goal is.

For democracies to work properly, power needs to be distributed, not concentrated, and all the reforms I suggest in this book work from that basic principle. They are designed to ensure that we-the-people are in a position to wield power, whether that is in government itself, in the workplace, or in the formal and informal spaces of civil society more generally. The goal is to build not just a fairer society, but a life in common in which we can all thrive, and that is why I will now turn to the concept of 'the commons'.

COMMONS

> Every time we take a breath, we're drawing from the commons. Every time we walk down a road we're using the commons. Every time we sit in the sunshine or shelter from the rain, listen to birdsong or shut our windows against the stench from a nearby oil refinery, we are engaging with the commons ... The commons make life possible.
>
> *Antonia Malchik,* Aeon Magazine

We live in a world of limits, beginning with the earth itself. It is a source of abundance, but it is finite. There is only so much gold, so much oil, so much scandium and yttrium, the rare-earth metals that make your smartphone smart. Even where resources regenerate themselves, there are limits to what we can take from them without risking their ability to renew: oceans can be overfished, forests over-logged, animals can be hunted to extinction, the air itself clogged with pollution. As humans, we too have limits: limits to our physical strength, our mental abilities, to our patience, our goodness, our wisdom. Sometimes our ability to understand ourselves seems the most finite resource of all, or at least, the most mysterious. What is less mysterious is that we are born into a world of cooperation and interdependence, that human babies – for a longer period than any other animal – depend on the care provided by others. We are not only helpless for a considerable period of time, but our brains need to increase their capacity from birth to adulthood by 100 000 times so that we can deal with the complexities of the world. A human life of individual independence and maturity is therefore entirely predicated on the help we receive from others. Our ability to function as individuals and a species is enabled by that interdependence. And that interdependence is enabled by the things we share in common.

The previous chapter set out a theory of power. This one is about values, about the way we imagine and manage a life in common.

Let's begin with the concept of public space. A democracy needs it in order to thrive because we invest public space and the things we own in common with meaning, a meaning that derives from how essential they are to our survival. Over time, across generations, public spaces like parks and town squares, beaches and riverbanks, forests and mountain tracks become meaningful and precious to us precisely because they are the places we share a life in common, even when there isn't another soul around. They are the places where we rehearse and enact private rituals of intimacy – holding hands, kissing, eating and drinking, throwing a ball, walking and talking – in a public way. We come together in these spaces to be ourselves with others, with strangers, and they thus embody the paradox of our individuality being formed and made meaningful by our relationship with others.

Without these spaces, we lose a sense of ourselves as a community and we diminish not just our ability but our desire to function as a society. If we do away with such spaces, if we enclose them and privatise them, make them subject to the property rights of an individual or corporation, we lose the transcendence they allow. You only have to think about the different way people act and feel in a public square compared to the way they act and feel in, say, the atrium of a major shopping mall. That atrium may indeed be a beautiful space, it may be crowded with people, but it is not a *public* space, and those people you are rubbing shoulders with are not citizens but customers. The space homogenises us, its architecture designed for profit rather than socialisation, and you only have to sit on the ground in a mall or a foodhall in order to see how long it will take for a private security guard to appear and move you along. These spaces do not liberate us – let alone our souls – they bring us under private control and surveillance, no matter how invisible that control and surveillance may be on a day-to-day basis. Simply replicating the 'look and feel' of a public space does not make it public. And when public spaces are lost, or made private, we lose one of the preconditions for a properly functioning democracy, for a properly functioning humanity.

How we cherish public spaces raises questions about the idea of a life in common: how do we maximise the benefits we all get from our finite resources without destroying the world and each other? How do we balance the competing needs of seven – or ten – billion people to honour their individual freedom while recognising their need to live together? These aren't just questions about economic management. They go to the heart of how we govern ourselves more generally, about how life itself is impossible unless we protect our natural environment.

You would think the blindingly obvious reality of the bounded nature of our planet would ensure that we approached the world, and our own place in it, with a certain level of respect and humility. But it turns out that for many of us, our desire for things such as power, pleasure and revenge, and our ability to turn a blind eye to the consequences of our actions, are pretty close to limitless. We will literally kill each other, and in immense numbers, rather than admit to certain limits. Even if we don't resort to murder or war, we will often change course only in the face of undeniable catastrophe, and even then we can be stupidly slow to react, as we are showing now with climate change. The same is true of the growing inequality that besets the developed world, which is particularly frustrating. It is not as if *that* is some natural disaster we stand powerless in the face of; it is a direct consequence of our own actions and decisions, and yet we find it incredibly hard to face and fix. So fundamental is this force in our life – our failure to accept the trade-off between finite resources and infinite desire – that we have developed entire fields of study that try and explain it: psychology, psychiatry, sociology, anthropology, political science, among others.

Arguably the most influential area of study that takes limitless desires and finite resources as its subject is economics. There was a time when the specialists that politicians called upon to help them make decisions were historians, but that moment passed sometime around the First World War. Now it is economists they turn to, and so it is their presumptions about the world that have helped shape it and pushed it in its current direction. 'Practical men who believe themselves to be quite exempt from any intellectual influence, are usually the slaves of some defunct economist', as the economist John Maynard Keynes once said. Economics examines how resources are allocated under conditions of scarcity, and economists make the basic assumption that we humans will try and maximise the benefits we receive. It's not a bad assumption, but like nearly everything else, it has its limits.

In wrestling with problems of scarcity, and applying certain theories on how to respond to it, economists – largely through their influence on politicians, who then bend and shape the theory to satisfy the real-world needs and wants of those they consider their constituents – have driven us to a particular set of policy prescriptions that have now come to the end of their useful life. The collective name for these policy prescriptions is neoliberalism. This is another controversial term, with endless arguments about its 'real' meaning and overall usefulness. But as I say in *Why the Future is Workless*, neoliberalism is the best collective description we have for the economic practice that has dominated in most countries of the world for the past 40 or so years. David Harvey in his book *A Brief History of Neoliberalism* explains it as 'a theory of political economic practices that proposes that human well-being can best be advanced by liberating individual entrepreneurial freedoms and skills within an institutional framework characterized by strong private property rights, free markets, and free trade'.

Neoliberalism has created a way of thinking about the world, and our place in it, that has diminished our individual freedom by reducing nearly everything we do – from running a business to being a student, hospital patient, citizen, even a parent – to a market transaction; that is, one that is decided by the use of money. It is a reductive philosophy, almost a mythology, that sucks the complexity out of human existence. In so doing, it commodifies activities that really aren't (or, more moralistically, shouldn't be) market exchanges at all (childcare, health, education, the power and water supplies). By defining individual freedom as our ability to make choices in a market – we can choose to have this job, this brand of toothpaste, that house – it answers the question of scarcity, or how we decide who gets what, by putting a price on nearly every choice that we make. The market therefore allocates not just scarce resources, but almost everything to which we can, allegedly, attach a monetary value. It is this mindset – where freedom equals choice in a marketplace – that is the moral prerequisite for the sort of inequality that currently plagues developed nations. If price is what determines who gets what, then inequality must follow.

Built into this underlying presumption is the idea of endless competition: that to increase our choices, and by definition enhance our freedom, we must compete with each other. An all-encompassing entrepreneurial mindset is therefore inculcated in us – students are literally referred to as customers or clients – and everything we do becomes transactional and governed by the logic of markets. It obscures other ways of thinking about the things we do, because we need to compete constantly to maintain any semblance of equality. And, because the neoliberal logic pervades everything we do, to withdraw from one competition is to withdraw from all. We are left having to choose between the benefits of the modern world – all embedded in market transactions – and an ascetic, survivalist type of existence as the only way to escape the neoliberal paradigm. As columnist Ben Tarnoff writes in *The Guardian*, 'in our new era of monopoly capitalism, consumer choice is a meaningless concept. Companies like Google and Facebook and Amazon dominate the digital sphere – you can't avoid them.' Under such circumstances, we are almost forced into a Unabomber-type logic, where all technology is seen as the enemy and the only 'sane' thing to do is to live alone in a cabin in the woods (until you become so isolated that you decide to start sending letter-bombs to people you consider symbolic of the madness you are trying to escape). More likely, you just shrug and do your best to survive in the 'normal' neoliberal world, hoping that it will collapse under the weight of its own shortcomings. Or perhaps you embrace the technology and become a fanboy; or maybe you project onto it the problems of the world: Facebook has destroyed journalism; Airbnb is destroying neighbourhoods; self-serve supermarket checkouts are the slippery slope to dystopia.

To challenge the pervasiveness of the neoliberal dispensation is not to argue for perfect equality or the sort of collectivism that levels all human endeavour to grey sameness. It is not even to argue against markets per se. But I am arguing that some things essential to our existence need to be removed from the logic of markets, to be returned to, or installed in, non-market circumstances and practices. We must find better ways of managing our scarce resources that don't create winners and losers on the scale that currently occurs based on market transactions alone. We must treat freedom as something we are all entitled to and recognise that that freedom – our very individuality – arises from the life we live in common. The paradox of this formulation is striking, but it is nicely summed up in the expression 'my freedom to swing my arms ends at the tip of your nose'. This idea of freedom necessarily implies a trade-off – no-one can ever get exactly what they want all the time – and it says that real freedom arises from the system that best maximises the ability of each individual within this constraint. To get these trade-offs right, we need to be able to measure the world we live in, and increasingly, economists concern themselves with data, as this defence of the profession in *Prospect Magazine* makes clear:

> We analyse data. Gigas and gigas of data on how much people work, which jobs they do, what they buy and what they eat, how they do in school and other aspects of human life. We do so for the UK and many other countries around the world – rich and poor … We analyse this data to understand how people make choices, because that determines how they respond to policies and how they interact. You can ask us about taxes, social mobility, inequality, crime, poverty alleviation, pensions, roads, sanitation, public safety, and, obviously, wine, beer and cider prices.

The question this begs, though, is, what if the data is wrong? Or rather, what if the data economists typically consider creates a distorted picture of the economy – and thus society – as a whole? This is precisely the criticism increasingly being levelled at the key economic metric of gross domestic product, or GDP. Not all economists think GDP is a problem – see, for instance, Diane Coyle's book, *GDP: A brief but affectionate history* – but even the OECD has said, 'If ever there was a controversial icon from the statistics world,

GDP is it. It measures income, but not equality, it measures growth, but not destruction, and it ignores values like social cohesion and the environment.'

The contradictions are highlighted by economist Lorenzo Fioramonti: 'Undesirable conditions like diseases, traffic jams, disasters, pollution and crime … trigger economic transactions … by requiring more medical treatment, anti-smog devices, higher insurance premiums and larger jails' and thus they 'count positively towards GDP' even though they 'are certainly not evidence of prosperity'. This means that a 'country that depletes its energy sources and destroys the environment to prop industrial output is seen as productive by GDP'. Most tellingly, he notes that 'GDP only counts transactions that occur within the formal economy, hence disregarding all economic activities that are informal, voluntary in nature and are performed within the household, thus driving societies to commercialize social life, reduce leisure and free time and support large corporate-driven industrialization'. This means GDP ignores work mainly done by women, a point well made by academic Kathi Weeks in her book *The Problem with Work*, and economist Marilyn Waring's seminal study *If Women Counted*.

Although it stands at the apex of measures we use to understand the economy, GDP is part of an older and more general trend to use statistics to describe the world we live in. Embedded in the word *statistics* is the thing it seeks to describe, and the history of stats is the history of the *state* trying to make itself visible to itself. In the same way that maps enabled us to grasp geography in a more manageable way, abstracting vast swathes of land – ultimately the whole world – into something we could hold in our hands, statistics abstracted the range of things that happened in those lands and made them comprehensible. William Davies points out in 'How statistics lost their power' (*The Guardian*), 'Organising numbers into rows and columns offered a powerful new way of displaying the attributes of a given society. Large, complex issues could now be surveyed simply by scanning the data laid out geometrically across a single page.' What's more:

> These innovations carried extraordinary potential for governments. By simplifying diverse populations down to specific indicators, and displaying them in suitable tables, governments could circumvent the need to acquire broader detailed local and historical insight … Regardless of whether a given nation had any common cultural identity, statisticians would assume some standard uniformity or, some might argue, impose that uniformity upon it.

The trouble is, there is no real way to avoid this uniformity. Any statistical measure is always going to be partial, and any more comprehensive approach (like a census that examines the whole population) is likely to be too expensive and unwieldy to run on a regular basis. The best we can hope for is to constantly revise the measures we use, understand their shortcomings, and try to plug the gaps.

GDP isn't the only metric that needs a rethink, but it is the one that deserves the most attention, and many conclude that rather than replacing GDP, we need to refine it, to move closer to measuring the 'good' and not just the 'more'. One way of doing this might be to use the real-time data we can collect through our various devices and other online and mobile activities to give us a better picture of how people are behaving. Economist Tara M Sinclair, writing for the Project Syndicate website ('Economic forecasts in the age of big data'), lists a range of new, real-time data sources and metrics that measure things such as unemployment and that help economists make predictions about future behaviour. The job-search company Indeed (where Sinclair works) uses real-time jobs data to see 'which sectors are attempting to recruit the most candidates', a process she calls 'a powerful economic indicator when evaluating the labor market'. The Billion Prices Project at MIT 'measures inflation using real-time data on online purchases from hundreds of retailers globally'. The Google Price Index does a similar thing, while Google Trends 'offers insights from Internet search data'. Other data-mining projects use social media sites to look for 'useful leading economic indicators, including the Twitter hashtag #NFPGuesses, a weekly aggregation of predictions about non-farm payroll gains'. Other examples are Zillow, 'an online real-estate service [that] collects information about home sales and mortgages', and SpaceKnow, which uses 'satellite imagery to track production'.

The logic of this approach is appealing. As Sinclair says, 'these newly available data reflect the real-time behavior of economic actors, revealing previously undetectable shifts in the economy'. In some ways, they overcome the partiality of other measures, and can therefore be less reductive. But we need to be careful.

The huge advantage traditional statistical practices and measures like GDP offer is their ability to abstract vast amounts of data into something we can easily run our eye over. Such measures give unity to the vast mess inherent in human life, especially at the level of the nation-state, and while something is lost in that process – key aspects of the diversity of human existence – something is gained in making that diversity manageable. Manageable in this sense, of course, implies being harnessed by the power of the state, and that, too, has its problems. With any state, no matter how ostensibly democratic, we always need to be wary of how centralising power – whether military or statistical – may flower into outright totalitarianism. It is, after all, no aberration that the Nazis kept meticulous records.

But you don't need to conjure totalitarianism in order to understand the shortcomings of the simplifications implied by GDP and other standard measures. Again, generalising, *by definition*, leaves out the specifics, and we come to think of the state or nation as embodying these generalisations. As William Davies points out, 'Headline-grabbing national indicators, such as GDP and inflation, conceal all sorts of localised gains and losses that are less commonly discussed by national politicians'. He says that 'Immigration may be good for the economy overall, but this does not mean that there are no local costs at all', and that 'when politicians use national indicators to make their case, they implicitly assume some spirit of patriotic mutual sacrifice on the part of voters'. But, he wonders, 'what if … the same city or region wins over and over again, while others always lose? On what principle of give and take is that justified?'

In Britain, those who support leaving the European Union tend to be part of certain demographics (older) and geographical areas (northern) and in no way represent the entire nation. Britain itself is an aggregation that barely holds together anymore as its constituent parts – England, Scotland, Wales, Northern Ireland – increasingly assert their independence. The people who wear red 'Make America Great Again' baseball caps are similarly geographically and demographically concentrated, again older and in the so-called flyover states (a term that underlines the weak bonds of the US nation-state). In Australia, as recently as 2017, there were calls for Western Australia to secede from the Federation, and there are ongoing regional differences in everything from sport to attitudes to immigration to Aboriginal rights that challenge the very notion of 'Australia' as a viable whole. To quote everyone's favourite poem, the centre cannot hold.

The backlash against experts, against the technocratic state based in the totalising power of statistical measurement, is often characterised as populism, but it goes much deeper than that and is reflected, for instance, in sophisticated critiques of capitalism and the state by the environmental movement. We are kidding ourselves if we think such a backlash is merely populist dissatisfaction being stirred up by demagogues. A reckoning is overdue. But if we're going to substitute big data for more traditional measures like GDP, it's important to realise that it comes with its own set of limitations. The key one is that most of the data is generated and interpreted by private organisations which, by and large, are seeking not to better explain the world to us, but to exploit it for profit. William Davies argues that 'the battle that will need to be waged in the long term is not between an elite-led politics of facts versus a populist politics of feeling' but 'between those still committed to public knowledge and public argument and those who profit from the ongoing disintegration of those things'.

This is not entirely convincing given that Davies has said that such disintegration is already happening with the emerging critiques of traditional measures like GDP. Indeed, many of those questioning the ongoing efficacy of the traditional measures are themselves former wielders of them, and so we have, for instance, the 2008 commission launched by then French President Nicolas Sarkozy to redesign GDP. He invited economists and Nobel laureates Joseph Stiglitz and Amartya Sen and others to run the project, and they produced a major report called *Mismeasuring Our Lives.* In it they state that, 'If we do not want our future and the future of our children and grandchildren to be riddled with financial, economic, social, and environmental disasters … we must change the way we live, consume, and produce … We will not change our behavior unless we change the ways we measure our economic performance.' In part as a response to the report, the Italian government in 2017 introduced new measures of national well-being, informally known as La Dolce Vita index (what else!). Bloomberg reported that, 'the finance ministry will produce official forecasts for 12 indicators, ranging from income inequality to carbon dioxide emissions to obesity – the first country to do so in the EU and the G-7'. New Zealand is in the process of introducing a similar measure to help determine their economic and social priorities.

There are other attempts underway to combat the privatisation of data and to instead collect and use it for public purposes. In 2017, the Australian National University announced that it had employed Genevieve Bell, a cultural anthropologist who had worked for Intel for the previous 20 years, to devise a new field of cross-disciplinary study. The idea is to use the new flows of data precisely in the way Davies suggests is necessary, with a commitment to public knowledge and public argument. Bell told the

Australian Financial Review, 'We're moving into a world which is much more driven by data and the circulation of data, the sense-making around data and things which move on the basis of data like algorithms, autonomous machinery, and artificial intelligence.' This means that we need 'a new body of knowledge and, as a result, a new way of thinking and operating'.

Encouraging work is also being done by organisations such as the Institute for New Economic Thinking at the Oxford Martin School. Its 2017 report *The Wealth of Nature* argues that although '[h]uman prosperity and wealth has increased dramatically over the last 200 years ... Societal risks arising from interruptions of supply chains, extreme weather, species losses and erosion of topsoil and reduced agricultural yields are being documented around the world.' To help combat these problems, the report suggests we invent new forms of measurement, noting that 'all natural capital – including minerals, resources, fossil fuels, but also valuable ecosystem assets and natural infrastructure – could support greater prosperity if it were more appropriately valued and hence more efficiently used'. Ultimately, the report argues, 'natural capital could be accurately reflected in comprehensive national wealth accounts, which serve as a better guide to economic progress than measures such as Gross Domestic Product (GDP)'.

We don't have to solve the problem of measurement more generally, or GDP in particular, right here, right now, but this gives us a sense that we are on the precipice of major changes about how we think about the world, including the way we measure what's important to us. What I am building up to here is a theory of the *commons* that offers us a way of understanding entities such as the state and the economy, one that better reflects the diversity that tends to be hidden from national indicators like GDP and counters the potential risks of privatised data flows. Davies is right to say that these new methods of data analysis are 'a form of aggregation suitable to a more fluid political age, in which not everything can be reliably referred back to some Enlightenment ideal of the nation state as guardian of the public interest'. But rather than seeing this as a wholly negative thing, we should exploit its potential by rethinking some of our presuppositions about the way the world works and then, as I do in the next section of the book, designing new public institutions, or radically redesigning old ones, that use these insights to better reflect and value we-the-people, in all our diversity, in this 'more fluid political age'.

In managing infinite needs and wants in a finite world, the presumptions we begin with matter, just like the data we use to give shape to the world matters. And this is, ultimately, why the idea of the commons matters too. According to the Digital Library of the Commons, 'the commons' is 'a general term for shared resources in which each stakeholder has an equal interest'. Typically, we think of it in terms of shared natural resources such land, air, water and minerals, but it also has a cultural dimension involving public stores of knowledge and the free exchange of ideas. Increasingly it is applied to the digital world of online data and other intangibles. I am using it to apply to all these situations, but also in a more metaphorical sense, as an idea that expresses what I take as a fundamental need of humans to cooperate in the use of shared, scarce resources.

James C Scott, in his book *Seeing Like a State*, makes a point that perhaps I haven't made strongly enough yet, which is that the totalising tendency states display when they impose measures like GDP is not necessarily a bad thing. 'The social simplifications thus introduced not only permitted a more finely tuned system of taxation and conscription but also greatly enhanced state capacity. They made possible quite discriminating interventions of every kind, such as public-health measures, political surveillance, and relief for the poor.' He makes another point that is equally important: 'large-scale capitalism is just as much an agency of homogenization, uniformity, grids, and heroic simplification as the state is ... Today, global capitalism is perhaps the most powerful force for homogenization, whereas the state may in some instances be the defender of local difference and variety.'

The failings of both the state and the market have led some to pursue an idea of the commons that relies on neither, or rather, that doesn't privilege one over the other. Chief among such theorists is economist and Nobel laureate Elinor Ostrom, whose book *Governing the Commons* is becoming increasingly influential. Her work tends to reach across left–right political divides, and you find praise for her from green activists as well as market libertarians. Nonetheless, the contemporary popularity of the concept of the commons has its origins in another writer, Garrett Hardin, whose 1968 essay 'The tragedy of the commons' entrenched it as not just a subject of study but as a paradigm-shaping concept. Unfortunately, Hardin's interpretation of the commons has proved inept, which we know largely through the work of Ostrom. Hardin declared that a common field was likely to be ruined by overuse as individual users pursued their own rational ends – by trying to graze as many animals as suits them, and them alone, in a given field (rational self-interest) – and the only way to overcome this tragedy was to allow individuals to own parts of the field outright, thus converting it from common to private property. Ostrom's painstaking empirical investigations over many decades shows example after example of people using a commons and avoiding the tragedy of overuse without resorting to private ownership. She shows conclusively that it is perfectly possible for sane humans with interests in common to manage a common resource like a field without ever descending into the tragedy that Hardin saw as inevitable.

In fairness to Hardin, his use of a common field was metaphorical. His real concern was overpopulation on a global scale, and he used the commons to illustrate what might happen if people just kept having children with no regard to the limits of the earth. As such he was eugenicist and held all sorts of other obnoxious beliefs (he believed in forced sterilisation and that poverty was natural, for instance), but he did, at least, end up acknowledging the shortcomings of his original analysis. In many ways 'The tragedy of the commons' can be seen as a founding document of the neoliberal revolution. Academic David Harvey says in *Rebel Cities* that he has 'lost count of the number of times I have seen Garrett Hardin's classic article … cited as an irrefutable argument for the superior efficiency of private property rights … and therefore an irrefutable justification for privatization'. It remains a powerful story for free marketeers to tell to justify the primacy, as they see it, of private property. It just happens to be wrong. Or, at least, it stands as a great example of how a potentially powerful insight can be over-generalised and thus blind us to other possibilities, in much the same way that GDP as a measure of national wealth hides as much as it reveals.

Ostrom's approach to the commons posits a form of ownership and control outside both the market and the state. She writes:

> one can observe in the world ... that neither the state nor the market is uniformly successful in enabling individuals to sustain long-term, productive use of natural resource systems. Further, communities of individuals have relied on institutions resembling neither the state nor the market to govern some resource systems with reasonable degrees of success over long periods of time.

She takes issue with those like Hardin who insist that privatising ownership is the only way to manage a commons, but she is equally convinced that simply substituting state for private control of the resources is just as misguided. She sees the privatisers and the advocates of state control as two sides of the same coin, both wishing to impose outside control, which is the principle she most strongly rejects. Indeed, much of her work stresses the need for solutions to be responsive to unique local conditions and to be flexible enough to change over time.

Her extensive research describes situations where common ownership of a resource (she typically examined waterways or fisheries or other natural resource areas) failed in ways similar to that specified by Hardin; but she found that this failure was far from the foregone conclusion Hardin had suggested. Looking at those situations where the management of the commons was handled successfully, Ostrom was able to extrapolate the 'rules' for that success, which Derek Wall lays out in his book *Elinor Ostrom's Rules for Radicals*. A sustainable commons needs 'clearly defined boundaries', to which 'commoners in a defined community had access'. This 'made it easier to reduce the problem of free riders', or those who tried to use the resource without paying for it or contributing to its management. As mentioned above, 'the rules for commons' use had to fit local circumstances', with those using the commons participating 'in the making and modifying of rules', because people are 'more likely to respect rules that they have helped construct'. These 'boundaries and effective rules will only work if they are policed in some way', with sanctions 'graded from soft to more severe'. Ostrom cites a Japanese example, where commoners were fined in sake which was then used to pay the constables who policed the commons. A commons, even one with simple rules, requires a system of 'low-cost conflict resolution', even if a highly informal one. It also needs a 'minimal recognition of rights to organise', as well as a wider system of 'negotiating the links between interlocking commons'. Notice how these rules of commons management reinforce the idea of power as dispersed and bottom-up, which I argued for in the previous chapter.

An obvious concern with Ostrom's work – if we want to apply its lessons more broadly – is that she looked at fairly small-scale examples. David Harvey points out that the largest commons Ostrom cites in *Governing the Commons* involved 15 000 people. Would her insights even apply to larger-scale examples? Eventually Ostrom herself thought they did, and she went on to investigate (with Charlotte Hess) the idea of the whole of human knowledge as a commons, which I will come back to. Harvey's concern is that in a small-scale commons, the management structure can be fairly flat, but that sort of horizontal arrangement will not work when applied to a larger commons problem:

> The possibilities for sensible management of common property resources that exist at one scale (such as shared water rights between one hundred farmers in a small river basin) do not and cannot carry over to problems such as global warming, or even to the regional diffusion of acid deposition from power stations.

I think Harvey's concerns are worth taking seriously, but are finally overstated. To some extent, they revolve around terminology, and he points out – disapprovingly – that Ostrom 'avoids' the word hierarchy, preferring to speak of 'nested' levels of control. While I suspect this simply reflects Ostrom's general approach of emphasising the cooperative nature of commons' management, Harvey is right in insisting that at scale, managing a commons *will* involve hierarchical control, and maybe that should be recognised in the terminology. He is right to be frustrated with left-wing activists who eschew any sort of hierarchy and insist on practices of horizontal decision-making that lead to no decisions being made. He is also right in pointing out that 'Questions of the commons ... are contradictory and therefore always contested', which means that 'the analyst is often left with a simple decision: Whose side are you on, whose common interests do you seek to protect, and by what means?' Answering that question involves making tough political decisions that, by necessity, imply hierarchical controls, particularly around the notion of enclosure. As Harvey says:

> In the grander scheme of things (and particularly at the global level), some sort of enclosure is often the best way to preserve certain kinds of valued commons. That sounds like, and is, a contradictory statement, but it reflects a truly contradictory situation … It will almost certainly require state authority to protect those commons against the philistine democracy of short-term moneyed interests ravaging the land … So not all forms of enclosure can be dismissed as bad by definition.

Ostrom's point that there must be 'ways of negotiating the links between interlocking commons' suggests the need for changing management practices once we scale up to bigger, even global, matters like climate change. None of that undermines the key reason for evoking the ideas of a commons in the first place; namely, to offer a way of thinking about how we manage these sorts of problems that doesn't reduce us to choosing between state and market options. I can illustrate this by looking at one of the most pressing matters of common ownership in modern democracies today: the so-called natural monopolies.

While we don't approach an electricity grid or a broadband network with the same sense of affection that we do a lake or a public square, there is a strong sense in most democracies that people need certain necessities of life to be under our control and ownership rather than in the hands of for-profit corporations. We want natural monopolies (or public or common goods) such as power, roads, railways, water and even communication, in public, not private, hands. Privatisation of public assets has been one of the most unpopular government policies of the last 40 years, and across democracies, opinion polls regularly show that most of us would like to renationalise these assets, to bring them back under common ownership. Pollsters Peter Lewis and Jackie Woods, writing for the ABC ('Hate privatisation? There's nothing new about that'), say that 'familiarity with privatisation [in Australia] has bred contempt in the electorate. Public views on privatisation are firmly negative and consistently so.' In Britain, research by YouGov suggests that not only is privatisation unpopular per se, but that majorities would like to see the Royal Mail (65 per cent) and the railways (60 per cent) renationalised. Eighty-four per cent of those polled want to keep the National Health Service in public ownership, and 81 per cent want schools to be publicly owned.

Support for public ownership is not merely sentimental. Had privatisation delivered the cost savings and efficiencies its advocates promised, then maybe people would be more accepting of the loss they feel. But on these measures alone, privatisation of natural monopolies has largely failed, and this failure compounds people's dissatisfaction with it, the feeling it brings of citizens losing control of their country. Until recently, renationalisation was considered impossible – that you couldn't unscramble that particular egg – but this feeling is rapidly fading. The British Labour Party, for instance, now has renationalisation of certain assets as party policy, while various economists and think tanks are investigating how it might be done. Writing in *The Guardian* ('We can undo privatisation. And it won't cost us a penny'), Will Hutton suggests the invention in Britain of 'a new category of company', which he calls the public benefit company (PBC). This new corporate structure 'would write into its constitution that its purpose is the delivery of public benefit to which profit-making is subordinate'. As well, the PBC would 'take a foundation share in each privatised utility as a condition of its licence to operate, requiring the utility to reincorporate as a public-benefit company'. As PBCs 'would remain owned by private shareholders, their borrowing would not be classed as public debt. The existing shareholders in the utility would remain shareholders, and their rights to votes and dividends would remain unimpaired.' This means governments would not need to buy back ownership and so the process of renationalisation would not cost the government – us – anything. We would therefore get the benefits of public control without having to buy out the current owners.

John Quiggin, professor of economics at the University of Queensland, suggests, instead, a structure that used to exist in Australia (before privatisation became preferred government policy) called the commercial statutory authority. Quiggin says this 'breaks with the idea that a corporate model, with directors responsible to shareholders is the best way to provide services to the public'. Instead, 'the public service objective is built into the governance of the organisation, rather than being imposed'. He points out that before privatisation, 'Australia Post was run, not by boards and CEOs but by public commissions, including representatives of customers, workers and the community at large, and charged with meeting "the social, industrial and commercial needs of the Australian people for postal services".' Its operating costs, over '15 years of operation as a statutory authority' were 'reduced by more than 30%, a reduction that compares favourably with the period since corporatisation'.

This is similar to the model being championed by John

McDonnell, Britain's shadow treasurer, who would like not just to renationalise these assets, but to bring them under *worker* control. He told a public meeting in London in February 2018, 'We should not try to recreate the nationalised industries of the past [and we] cannot be nostalgic for a model whose management was often too distant, too bureaucratic'. Rather, he suggests developing a different kind of public ownership based on the idea that 'nobody knows better how to run these industries than those who spend their lives with them'. Seen this way, such cooperatives are about much more than simply who runs and owns a given industry; they are about equality and ownership. Responding to McDonnell's proposal on openDemocracy ('The new economics of Labour'), Hilary Wainwright notes that it represents not just a way of renationalising these industries, but 'a new and very different understanding of knowledge – even of what counts as knowledge – in public administration, and hence of whose knowledge matters'. Wainwright says that for 'industries to be run by "those who spend their lives with them" means recognising the knowledge drawn from practical experience, which is often tacit rather than codified: an understanding of expertise that opens decision-making to wider popular participation, beyond the private boss or the state bureaucrat'. The very structure of this sort of cooperative is a way of distributing power in the way described in the last chapter, and as such, institutionalises democratic control in the fabric of the enterprise. This is a practical application of Ostrom's ideas about the commons at scale.

And there is another important reason for taking such action. Privatisation fuels the overall sense that government is ignoring the will of the people and is more interested in appeasing various business interests. Bringing such assets back into public ownership is a powerful signal that governments are there to serve the common good. It gives people a sense of control similar to that engendered by being able to access public spaces, and is part and parcel of what it means to live in a democracy. I would go so far as to say that the renationalisation of these natural monopolies is also a necessary precondition for popular acceptance of globalisation. 'Globalisation' has become a bogeyman in most democracies, which is unfortunate given the benefits of international trade, of more open borders and of recognising the need for a common response to problems that transcend national borders, such as climate change. A life in common is meaningless without a planetary perspective, but that perspective needs also to honour people's attachment to *place*, to home. Renationalising natural monopolies helps get the balance right between realising the benefits of an internationalised world and the need for people to feel in control of their own destinies in the places where they live. John McDonnell is right to say that renationalisation of natural monopolies is of benefit, 'not only because it's the most efficient way of running them, but also because the most important protection of our public services for the long term is for everyone to have and feel ownership of them'.

If we renationalise natural monopolies via something like the statutory authority model or a cooperative, we are baking the notion of common ownership and control right into the walls of the governing structure. We are modelling the sort of non-state, non-market model that I think is necessary to realise a life in common, everything from the public ownership of natural assets to the notion of a genuinely sharing economy.

Of course, simply citing the idea of a commons does not absolve us from the hard grind of political activism – of politics itself – that I discussed in the previous chapter, and that is not the point of it. As Hess and Ostrom say, a commons does not guarantee a positive result 'and its outcome can be good or bad, sustainable or not – which is why we need understanding and clarity, skilled decision-making abilities, and cooperative management'. Instead, the commons is offered as a way of breaking out of the business-as-usual politics that we are all so fed up with and that is proving useless in addressing the problems that confront us. Just as changing how we measure national wealth will force us to think differently about the way we run the economy, changing how we think about other social structures will bust us out of old ways of doing things. As David Bollier notes in the chapter he contributed to *Understanding Knowledge as a Commons*, 'To talk about the commons is to assume a more holistic vantage point for assessing how a resource may be best managed.'

To evoke the commons is to insist on a view of human life that tries to better negotiate the claims of the individual and the needs of the many. It is to recognise our need to live a life in common to properly realise our individual potential. It is to dethrone the idea that the only way of running things efficiently is through centralised control. It is also about recognising a bottom-up way of organising power, and it is no coincidence that the principles of organising I outlined in the previous chapter loom large in the theory and practices of the commons. In the following chapters, we will see how these underlying principles – of wielding power and managing the commons – might be applied in the areas of media, government, wealth distribution, work and education.

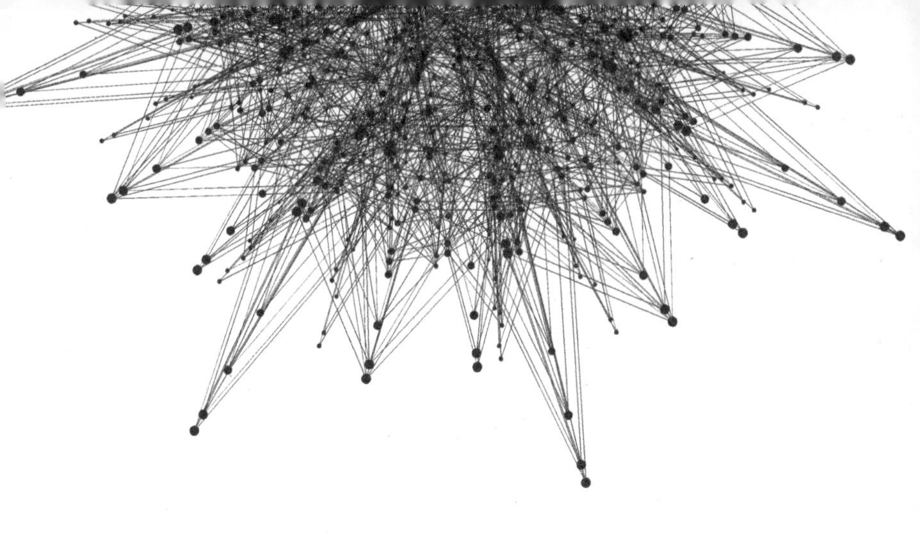

PRACTICE

MEDIA

> If we want things to stay as they are, everything will have to change.
>
> *Tancredi, in Giuseppe Tomasi di Lampedusa's* The Leopard

A free press has always been at the heart of democracy because we understand that without an independent institution able to question the state itself, the state becomes too powerful: power is concentrated rather than dispersed and we risk tipping over into tyranny. If democracy is self-rule, then the rulers – the citizens – need to be free to speak independently of any government they create. The traditional mainstream media became the institutional embodiment of this idea, and in the same way that we delegate control of government to elected representatives, we delegate the watchdog role over those representatives to a professional class of political journalists whose job it is to monitor and report back to us on how we are being governed. When Walter Cronkite went on television in 1968 and told viewers that American soldiers in Vietnam were 'mired in stalemate' and not, as the government insisted, close to victory, he was not just acting as a journalist reporting the facts, but as a steward of democracy itself. (And just to be clear, when I am speaking about the media in this chapter, by and large I mean that part of the media that covers politics.)

The advantage of having a purpose-built institution watching over government is that it not only frees up citizens to get on with other things, but the institution itself becomes powerful and has the ability to go toe-to-toe with other sources of power – the state, corporations, various elites – in a way that the individual citizen cannot. The disadvantage is that the representative nature of both government and the media not only puts a distance between we-the-people and our elected representatives, but it brings the politicians and the media into a symbiotic relationship. The government relies on the media to explain to the rest of us how it is choosing to govern, and the media relies on the politicians for access to the information they need to report back to us. So although at heart the relationship between media and politics is adversarial, it actually requires a high degree of cooperation and coordination between the two institutions. This inevitably leads to a degree of capture, of group sympathy, which is why I think it is correct to think of the press and the politicians as part of the political class.

There is another aspect to this capture. Mainstream media organisations are themselves, by and large, private businesses run for profit. This means that the media has a degree of self-interest in how government is run, and the temptation to influence government – by the promise or practice of favourable coverage – is irresistible. Traditional media thus exist in the fraught position of being part of the very power structures they are supposed to be reporting on, and it is hardly surprising that self-interest often outweighs a commitment to fearless exposure. Indeed, the media cling to the idea of objectivity – that they are above the fray and 'just reporting' – precisely as a way of holding the line against the pull of self-interest. Objectivity, balance, fact-checking and the like are part of the professional toolkit the media rely on to justify their claims of being honest brokers between we-the-people and the sources of power they report on.

All this means that 'Who watches the watchdog?' becomes a natural and vital question. Who actually decides whether the media are living up to their own standards? The answer to that question is: *we* do. So if we are unhappy with the way we are being governed, if we think politicians are failing in their basic duties of representation and collective governance, we will likely also think the media is failing. And boy, have we ever reached that point. This isn't necessarily a bad thing, but it can be dangerous, especially if the media itself is reluctant to acknowledge its failings (and yes, we have reached that point too). What's worse, when the watchdog is suspect, a politician with an eye for the main chance can exploit our dissatisfaction with the media: never underestimate the power of having a good enemy.

No politician has exploited this idea more shamelessly than US president Donald Trump. Trump thrives on creating enemies, on bullying almost anyone, and in that sense, journalists were an inevitable target. What was less obvious was the extent to which in going after the media, he was pushing an open door. Trump has positioned himself to be able to ignore or undermine the media's ability to hold him accountable by capitalising on their general unpopularity, and – to date, at least – it has worked very well: hatred and mistrust of the media crosses party-political lines. As the industry itself has been financially undermined by the rise of social media platforms, and as its audiences have begun using those same platforms to express *their* views on how journalists do their jobs, it has become undeniably apparent that the media have few friends. Journalists try and spin this as a positive by saying things like, 'If everybody hates us, we must be doing something right', but that is a weak defence, as believable as saying 'Sticks and stones will break my bones, but names will never hurt me', an incantation chanted *only* when the names begin to hurt. When Trump started attacking the media, accusing them of bias and being peddlers of fake news, he had found an enemy already in crisis.

Trump's rise has added some urgency and focus to criticisms of the media. As destabilising as his attacks have been for the press, they have also reminded people what a cornerstone of democratic government the media can be, and this is reflected in people not only watching, reading and sharing more news stories, but in increasing the amount of money they spend on them. For years, audiences were reluctant to pay directly for news, but there is some evidence that is changing. *The New York Times* added more subscribers in the last three months of 2016, around the time of Trump's election, than it did in all of 2015. Even a non-political magazine like *Vanity Fair* had its biggest single-day increase in subscriptions after Trump tweeted that the magazine's circulation was 'way down, big trouble, dead!' *The Guardian*, which has struggled for years, and has been dipping into the capital component of its trust-fund wealth (rather than living on the interest, as it is meant to do), announced at the end of 2017 that it was 'on track to … break even … and for the Guardian Media Group to become financially sustainable in perpetuity'. This success was built on the back of a massive increase in reader membership, with over 400 000 paying members now on the books.

So let's try and get our heads around this. If the political media are failing us, and we want to rethink the way they do their job, we need to be honest. We need to understand why the door Trump pushed was already open. The media as an industry glories in its role as a fourth estate, as a check on power, 'afflicting the comfortable and comforting the afflicted', and as one of the key pillars of a functioning democracy – and it is those things – but it has always been, as well, a cesspool of inaccuracy, bullying, titillation and sensationalism. Corporate media are often a functionary of the state and the elite interests that control them, and they are just as likely to side with power as oppose it. This is particularly true in times of war, when the 'free press' can be counted on to rally support and manufacture rationales for even the dumbest interventions. For every Watergate exposure, there are ten other examples where the media backed off or willingly decided to comfort the comfortable or afflict the afflicted. You would expect politicians to have a low opinion of the media and to feel aggrieved when they are held to account, but when I wrote my book about media, *The New Front Page*, I found the same concerns from those whose contact with the media was more sporadic. I spoke to person after person who had dealt with a journalist – often for the first time – and who said they had been taken out of context or even betrayed. They might've been contacted for an expert opinion or a quote, or attracted media interest for some other reason, but they all voiced the same complaint: they were misrepresented by a journalist who didn't understand or care what they were saying and who shoehorned their comments into a preconceived narrative.

Nicholas Gruen, the former chair of the Australian Centre for Social Innovation, which has helped run citizens' juries in South Australia, recounted to an audience at the 2016 Adelaide Festival of Ideas comments from the citizens involved in the juries. Participants were overwhelmingly positive about their experience, except, Gruen noted, in one regard. 'There was only one institution which people began with a low opinion of, but whose reputation sank further as jurors learned more. The media.' A citizen juror said, 'It was eye opening, because you had your opinions about things, and you would hear something completely opposite … and then you realise that you never really get the full story through the media.' Numerous jurors, Gruen said, were 'angry', 'annoyed' and 'appalled' when they saw media coverage of their deliberations.

Those of us involved in the early days of blogging had a similar reaction to the way in which the build-up to and execution of the US invasion of Iraq were handled by the media. Bloggers like me were pretty well educated in the problems of corporate media, we understood how facts could be manipulated, but the rigours of daily blogging, which have you poring over the media's statements in a way that most of us don't, were eye-opening. 'Iraq proved to be a turning point', I write in *The New Front Page*:

> Not only did much of the mainstream media fail to see behind the false case presented by the Bush administration for the invasion, but they were also shown to have failed in that basic duty by the so-called new media, the endless stream of blogs that blossomed in the lead-up to and prosecution of that war. If the mainstream media could get the reporting wrong on

> something as important as a war, when mere amateurs could see through the nonsense, what else was it getting wrong? That was the question a key segment of the audience suddenly asked themselves, and they've been asking it daily ever since.

It is not enough to say, as many do, that such failings do not apply to every aspect of the media, or even that it doesn't apply to every journalist. These statements completely gloss over the fact that the problems are structural, built into the way the job is done. Asserting counter examples, the good things journalists do, the telling and compelling stories they write, does nothing to fix the underlying problems. The toolkit is broken and needs to be replaced. Until it is, journalists will continue to do bad work.

For a brief, shiny moment at the beginning of the new century, some of us thought that the rise of 'new media' might help fix these underlying problems, but we are far enough along now to say that it has merely changed the terrain in which they occur. In many ways, new media has amplified existing bad habits. Don't get me wrong: I wouldn't for a second go back to the pre-internet days of conglomerate media and an audience restricted to the letters pages or the phone queues of talkback radio, but the shift online has not been the democratic vitamin that some (like me) had hoped for. Or rather, it has, but it has also bred a new and powerful strain of anti-democracy, and the flow of these two forces through the bloodstream of online news is defining the new era.

We are still in the habit of talking about new media and old media, but I think it is time we broke that habit. 'Old media' was the period of centralised control of the news and the dominance of a number of big newspapers, television and radio stations. It is over. 'New media' describes the transformation that occurred when traditional newsgathering services began to digitise their content and make it available online. It describes the period of disaggregation, of an explosion of new forms – everything from the blog to the meme, to the podcast, to the listicle – and the rise of the audience. But the era of new media is over too. We have now entered a period I call 'fusion media', in which new media platforms such as Google, Facebook and Twitter coexist with the old media of television, radio and newspapers. Journalism happens in a hybridised environment, where traditional reporting mingles online with the actions and habits of an audience able themselves to shape, distribute and create content. This new space brings together amateurs and professionals in a way that wasn't previously possible and creates new and different relationships of power between the media and its audience and, most importantly, between the media and politics.

So what's the difference between new and fusion media? The dates are somewhat arbitrary, but I suggest new media began in 1980, when 12 newspapers that were members of the Associated Press – including *The Washington Post*, *The San Francisco Chronicle* and *The St Louis Post-Dispatch* – went online via the CompuServe dial-up service. Back then, only about 20 000 people in the world had access to the internet, via universities, and military and government departments, but this was the first serious attempt to make journalism available online. The period continued into the late '90s and early 2000s as more and more traditional news companies came online, and as the rise of personal websites and blogging platforms allowed the 'people formerly known as the audience' (to use journalism professor Jay Rosen's phrase) to inject themselves into this new-media space and to use other online technologies – particularly the Google search engine – to challenge the primacy of the traditional media's role as interpreter of the political landscape.

I date the moment when new media gave way to fusion media from the launch of Facebook in February 2004. Although blogs themselves were a form of social media, and we had the period of MySpace and GeoCities and other more amateurish attempts to 'get online', it was the rise of these new platforms, particularly Facebook and then Twitter two years later, in March 2006, that really defines the fusion media period. Their existence signalled the end of the somewhat anarchic phase of the internet. The deliberative space created around the network power of blogs – their ability to link to each other and to other sources of information – shifted and changed in significant ways. The rise of algorithms – particularly Facebook's – as a way of sorting what news is seen and distributed helps define the difference between new media and fusion media, as does the advent of mobile technologies.

The shift from old to new media made the symbiotic relationship between the media and politicians visible in a way that wasn't previously possible. And it changed the audience's role from passive observers to active participants; the media went from being a representative institution to a participative one. It was a new frontier, a Wild West, where people showed up and involved themselves in reporting and analysis without having to seek the permission of some editor or other gatekeeper. Fan sites and online diaries, discussion boards and email lists all sprang up like gold claims along the river in *Deadwood*. It was rule-free and radically amateur, and the strength of a lot of the political commentary that arose – especially when blogging took off in the early 2000s – was that the citizens involved didn't have to play by the rules of the corporate media, of professional journalists. This, of course, made them a target of those journalists, who mocked their lack of professionalism, their lack of objectivity and their lack of access, but that was the whole point. The people involved in these discussions had no interest in the rules of the game played by journalists and no interest in the prestige, power or even the income that drove the professionals. They had no need to maintain access with politicians and so no need to pull their punches. This was the joy of participation, of finding like-minded friends, or even those who disagreed with you vehemently, and being free to slug it out.

As such, the period of new media was a better deliberative space than it is often given credit for. It wasn't just that it injected the public into the conversation in a way not previously possible, but that it allowed the emergence of voices previously stifled by the guardians of the mainstream media. It brought a diversity that exposed the mainstream for the middle-class 'sausage fest' it tended to be. Although, like any technology, the internet reproduced the inequalities of the societies in which it became available, it also liberated many voices that would otherwise not have had an outlet. New media challenged the authority of the mainstream media while at the same time destroying the basis of its financial viability. That's a hell of a double whammy and the mainstream is yet to recover.

During this Wild West phase, many of us were willing to work for free. The internet was a commons to which we all contributed and from which we extracted non-monetary value. We built sites and wrote content and gathered audiences, and we argued and discussed and formed tribes, and none of us expected a career, advancement or a wage from these activities. Because of that, our audiences didn't have to pay either. We were in it as citizens, not entrepreneurs and consumers, and it all worked as long as we had other sources of income. Ultimately, some of us professionalised and were hired by media companies (this happened to me), or we started our own businesses by finding investors and setting ourselves up to compete commercially. But the logic of 'free' was built into everyone's practices and expectations: 'information wants to be free' chanted the aging hippies reborn as tech libertarians.

The trouble with 'free' in this sense is that, in a market economy, someone has to pay, and when the free model came to be supported by advertising, so the Wild West came to reproduce the offline world it was rapidly replacing. Advertising had supported most traditional media, with a given newspaper, television or radio station able to charge a price predicated on the audience size it could deliver to advertisers. That, in turn, was based on the ability of news outlets to dominate a particular geographical area, normally a city, and so when the internet rendered geography irrelevant – anyone could access news from anywhere – this business model collapsed. The advantage of geographical monopoly was replaced by the advantage of controlling online entry itself. As business analyst Ben Thompson notes, 'The fundamental nature of the Internet is abundance; its critical competency is discovery.' And because so many people now use either Google or Facebook as their primary way onto the internet – we either search for something or check in on our timeline – those two companies have been able to dominate discovery and therefore the revenue generated through online advertising. Traditional media organisations have become the poor relations in an online world as they compete with those two key choke points of internet entry. In 2016, Facebook and Google accounted for 99 per cent of the growth in revenue from online advertising, and more than 80 per cent of the total revenue it generates, statistics that exemplify the ongoing struggle media companies face for viability.

This has also meant the Facebook algorithm and Twitter's chronological timeline have become the preferred way for filtering the abundance of the 24/7 news cycle. Instead of bookmarking preferred sites and checking them individually, you receive content as the people you 'follow' on Twitter and 'friend' on Facebook share the things they are reading or find interesting. The platforms curate the news in the way that was once the sole prerogative of news editors. What's more, the discussions that once happened on blogs and in comments threads appended to articles in mainstream outlets have migrated to Twitter and Facebook. When this first happened, most people presumed that the Wild West aspect of new media would simply continue – after all, Facebook doesn't seem that different to a blog, and Twitter is often called a microblogging platform – and that they would just be able to better manage their involvement in it. The success of these platforms, therefore, was built on the expectations established by new media practices, even as the platforms betrayed them. As Wild Bill Hickok says to his friend Seth Bullock in episode 3 of *Deadwood*, 'Pretty quick you'll have laws here and every other damn thing', and so 'better management' came at a cost: the Wild West was tamed. The idiosyncrasy of blogging has become the uniformity of the platforms.

Mainstream outlets, too, have become more dependent on the platforms, because that's where the audience is. To be discovered among the abundance is a preoccupation of media companies (and individuals) who rely on this 'attention economy'. The 'Kremlinology' of the Facebook algorithm – figuring out what sort of content will suit its parameters – has become an industry in itself, and it is apparent that social media in general and Facebook in particular are affecting not just journalism, but our whole lives. In the *Columbia Journalism Review*, Emily Bell writes:

> Social media hasn't just swallowed journalism, it has swallowed everything. It has swallowed political campaigns, banking systems, personal histories, the leisure industry, retail, even government and security. The phone in our pocket is our portal to the world. I think in many ways this heralds enormously exciting opportunities for education, information, and connection, but it brings with it a host of contingent existential risks.

Media sites gather audiences and sell them to advertisers, and so content – news – is endlessly manipulated and shaped in order to generate hits, a game that Google and Facebook play better than anyone. What's more, the data that audiences provide simply by using the platforms is a vast source of information about them, their likes and dislikes, and an immense store of other personal information that the platforms can distil to help advertisers construct highly targeted advertisements, which are much more effective than the hit-or-miss display advertising that the media companies were stuck with. We have reached a point where the very raison d'être of the platforms is to crunch user data in order to maximise advertising revenue, or as former Facebook engineer Jeff Hammerbacher, puts it, 'The best minds of my generation are thinking about how to make people click ads.' The services Google and Facebook and other platforms provide are free (provided you can afford to be online in the first place), but they are paid for by advertising and so devalue any content that doesn't generate income in that way.

And this is the nature of fusion media. It has a whiff of new-media freedom about it – you can still log on for free and use the platforms for nothing – but, in fact, the commons has been enclosed. Instead of being a negotiated space of almost limitless participation and amateurism, the neoliberal logic of entrepreneurialism and professionalism now dominates. The platforms have swallowed all the content and assimilated it to their commercial architecture. That architecture includes everything from the character limits imposed by Twitter – which began at 140 and doubled to 280 at the end of 2017 – to the way in which the Facebook algorithm prioritises content. It includes the way content creators have to tweak that content so that it is better recognised by the Google algorithm ('search engine optimisation' and other techniques). It is an economy of attention measured in clicks and likes and retweets. Whereas once publication was built around the time demands of the morning or afternoon edition of the newspaper, the six o'clock television news, or the hourly radio bulletin, news is now ubiquitous, existing in a global online environment that never rests.

And if clicks and likes and the other reductive interactions aren't enough to suggest that participation has been dumbed down and hollowed out, fusion media – Facebook in particular – is an increasingly infantilised space of 'wellness' and 'lifestyle', where the corporate sheriff insulates us from anything that might upset us and cause us not to click on the ads. Announcing changes to Facebook's algorithm at the beginning of 2018, CEO and founder Mark Zuckerberg went the full lifestyle guru and told us, 'We feel a responsibility to make sure our services aren't just fun to use, but also good for people's well-being.' Which means less news from the real world that might upset us. (Former Australian prime minister Paul Keating once said, 'We don't want to be sparkling and happy all the time. You need the inner life, the inner sadness. It is what fills you out. There is a place for sadness and melancholy.' The very thought is anathema to Facebook.) Facebook also limits the ability of users to link to content outside the Facebook environment, so everyone's page looks the same. The radicalism of new media has transformed into the conformity of fusion media, which has a neutralising effect on the ability of people to use the platforms for democratic reform. As academic Wendy Brown says (in an interview with Tocqueville 21): 'Consider, for example, how many left intellectuals use their social media profiles – Twitter, Facebook, etc – not to build the Revolution, but to promote their books, speaking gigs, and ideas in order to boost their market value.'

What's more, the scandal around the way in which the firm Cambridge Analytica was able to access the data of Facebook users through the use of third-party apps (such as quizzes) has shown that Facebook's commitment to making the site 'good for people's well-being' was partial, to say the least. Although the full ramifications of these revelations are not yet known, they have at least exposed the extent to which social media platforms collect and store personal data, and the need for them to take a more responsible approach to people's privacy. This may involve more government regulation, but perhaps a better way to address the matter is for users to retain clearer ownership of their personal data and thus force companies like Facebook to compensate us for their access to it. I discuss this more fully below.

The media, as commercial players in a field over which they have limited control, must remain highly sensitive to how the platforms change and must adapt in order to ensure their own commercial viability. Having spent the best part of a decade mastering the vagaries of the Facebook algorithm in order to maximise the 'organic reach' of their content (the number of people they could reach without paying to 'boost' that content), media outlets at the beginning of 2018 were confronted with an announcement from Facebook that it would no longer prioritise news, that the algorithm would prefer the content that ordinary users created, rather than that created by brands and media companies. The point is not just the nature of the change but the fact that fusion media is an unpredictable space for media outlets, and they are always vulnerable to shifts beyond their control. The other key aspect of the fusion space is the way it deprofessionalises journalism itself, robbing journalists of the prestige and power that comes from exclusive access to information and publication, and allowing 'amateurs' who can use the platforms to command attention for *their* content to compete with the mainstream media's interpretation of any given event.

The 'fake news' phenomenon arises in these circumstances. It is both a product of the architecture of fusion media – the way it creates, constructs and distributes content – *and* a panicked response by media outlets in a constant state of flux in that space. Indeed, that panic is itself a feature of fusion media, reflected in the endless hand-wringing in the mainstream about every aspect of social media, from the way it infringes on 'their' territory, to demands from the likes of Rupert Murdoch that Google and Facebook pay media outlets for their content. Although Donald Trump has extended the term 'fake news' into meaninglessness, as he and his supporters use it to describe any news report they don't like, it is also used in the sense of false information circulated via social media. By optimising content to be picked up and distributed by various social media platforms, the argument goes, people with the intent of spreading false information, or even just confusion, are able to inject falsehoods into people's social media timelines. The practice is attributed to individuals and organisations acting on behalf of particular politicians, but it is also seen as part of larger operations by states to spread disinformation, the most commonly cited example of the latter being the alleged interference by Russia in the 2016 US presidential election.

Detailed examinations of the phenomenon by the National Bureau of Economic Research ('Social media and fake news in the 2016 election') and by researchers Yochai Benkler, Robert Faris, Hal Roberts and Ethan Zuckerman ('Breitbart-led right-wing media ecosystem altered broader media agenda', published in the *Columbia Journalism Review*), among others, suggest that 'fake news' wasn't as big a problem as the mainstream media tended to suggest, and what's more, the mainstream itself was likely more responsible for misinformation spread during the 2016 US presidential election than anything generated by social media.

This in turn bears out research by Andrew Chadwick (in his book *The Hybrid Media System*) that fusion media, paradoxically, is still dominated by the 'logics' of traditional broadcasting and newspaper media, 'who are in many respects successfully co-opting newer media logics for their own purposes, while at the same time restating and renewing the logics that sustained their dominance throughout the twentieth century'. He talks about the way in which 'the broad and continuing power of the political and media elites' has allowed them to carve out 'domains that enable them to control what are still the main vehicles for politics in a liberal democracy: organized parties, candidates' campaigns, and of course the extremely powerful, and increasingly renewed, mass medium of television'.

This means that while the ability of malicious sources to spread disinformation through social media platforms is a major issue, the mainstream media still plays an influential role in the way we-the-people are exposed to the political process. Fusion media may be where we experience such disinformation, but that doesn't mean it is generated exclusively by the new platforms. The *Columbia Journalism Review* (*CJR*), drawing on the sources I've mentioned above, suggests that the number of 'fake' Facebook ads purchased during the 2016 presidential election was vanishingly small, with 3000 ads generating around $100 000 in revenue. Put that in perspective with Facebook's overall user base and site revenue, and you get a sense of its relative impact: 'Facebook's advertising revenue in the fourth quarter of 2016 was $8.8 billion, or $96 million per day. All together, the fake ads accounted for roughly 0.1 percent of Facebook's daily advertising revenue.' A report by BuzzFeed into the matter found that 'the top 20 fake news stories on

Facebook "generated 8 711 000 shares, reactions, and comments" between August 1 and Election Day'. As *CJR* says:

> this sounds like a large number until it's put into perspective: Facebook had well over 1.5 billion active monthly users in 2016. If each user took only a single action per day on average (likely an underestimate), then throughout those 100 days prior to the election, the 20 stories in BuzzFeed's study would have accounted for only 0.006 percent of user actions.

In early 2018, Twitter released internal data showing the extent of the 'fake news' infiltration on its site. Most of the concern centred on an agency of the Russian government called the Internet Research Agency, which was known to be using Twitter to distribute false information during the presidential election campaign. However, Twitter reports on its company blog that its investigation uncovered 'a total of 50 258 automated accounts that we identified as Russian-linked and Tweeting election-related content during the election period, representing approximately two one-hundredths of a percent (0.016%) of the total accounts on Twitter at the time'.

Although there are arguments about how well this research reveals the extent of the problem, no-one suggests that it is orders of magnitude larger than outlined here (though information continues to emerge and the final story is yet to be written). To this point, however, *CJR* suggests, 'by any reasonable metric – including Facebook or Twitter shares, but also referrals from other media sites, number of published stories, etc – the media ecosystem remains dominated by conventional … sources such as

The Washington Post, The New York Times, HuffPost, CNN, and *Politico*.' During the same period in which the 'fake news' stories were detected, the researchers found that '*The Washington Post* produced more than 50 000 stories ... while *The New York Times, CNN*, and *Huffington Post* each published more than 30 000 stories.' And it wasn't just the quantity of stories that mattered. The research shows that these mainstream sources were skewed in ways that make a laughing stock of any suggestion that they are objective. On a sentence-by-sentence analysis, the results are quite shocking:

> They found roughly four times as many Clinton-related sentences that described scandals as opposed to policies, whereas Trump-related sentences were one-and-a-half times as likely to be about policy as scandal. Given the sheer number of scandals in which Trump was implicated ... it is striking that the media devoted more attention to his policies than to his personal failings. Even more striking, the various Clinton-related email scandals – her use of a private email server while secretary of state, as well as the DNC and John Podesta hacks – accounted for more sentences than all of Trump's scandals combined (65 000 vs 40 000) and more than twice as many as were devoted to all of her policy positions.

As *CJR* itself says, 'To the extent that voters mistrusted Hillary Clinton ... these numbers suggest their views were influenced more by mainstream news sources than by fake news.' The article concludes unequivocally:

> In sheer numerical terms, the information to which voters were exposed during the election campaign was overwhelmingly produced not by fake news sites or even by alt-right media sources, but by household names like *The New York Times*, *The Washington Post*, and *CNN*. Without discounting the role played by malicious Russian hackers and naïve tech executives, we believe that fixing the information ecosystem is at least as much about improving the real news as it is about stopping the fake stuff.

The question is, can we do either?

The matter of fake news cannot be separated from the rise of Donald Trump, and the key issue with Trump is not that he has chosen to attack the media, but that the media have been so vulnerable to the sort of attacks he has launched. The media have long relied on certain norms of political reporting: pitting opposing views against each other to generate controversy rather than enlightenment; providing false balance, where any view expressed has to be set against a countervailing view, no matter how marginal that countervailing view is; and 'backgrounding' or using off-the-record sources. These norms have always traded accuracy and analysis for access and exclusivity, nuance for timeliness, self-reflection for a misleading self-image of fearless reporting, and in so doing left the media vulnerable to manipulation. Trump's willingness to ignore them has definitely created a crisis for the media, but it was a crisis a long time in the making.

So the need for renewal goes much deeper than the current occupant of the White House, and unless the media realise that, Trump, once he has left office, will be dismissed as an aberration, and the structural problems will remain. It is telling that when one journalist, Michael Wolff, found a way to penetrate Trump's manipulation, specifically by ignoring the norms of conventional journalistic practice, many journalists attacked him and declared his approach poor form. Now, I am of the view that Wolff's approach could not be reproduced by the political media en masse, but it is indicative of the rethink that has to happen. Instead of criticising his approach, journalists should be refining it.

What Trump *hasn't* done is break out of the symbiotic relationship between media and politician. In fact, he has heightened it. He is desperate for the attention they give him, while they revel in the attention he brings to their publications, and so the media have become captive to his sensation-generating machine, apparently unaware the problem lies in their reliance on sensation in the first place. Les Moonves, chair of the CBS television network, inadvertently summed up the cravenness of the interdependence between then candidate Trump and the journalists covering him when he said, 'It may not be good for America, but it's damn good for CBS … Man, who would have expected the ride we're all having right now? … The money's rolling in and this is fun.' He could barely contain himself: 'I've never seen anything like this, and this going to be a very good year for us. Sorry. It's a terrible thing to say. But, bring it on, Donald. Keep going.'

His comments are a useful reminder of how the mainstream thinks – at its highest levels, anyway – and why we don't want to go back to the days when coverage, analysis and commentary of politics was the preserve of traditional media alone. But nostalgia about new media is not something we want to cultivate either. The freedom the Wild

West offered was illusory, and in its ungoverned form it was nearly overrun by thugs and gunslingers who wanted to dominate a fiefdom rather than share a commons. We are all familiar with the abuse and vitriol that became part and parcel of many online forums, congealing into organised abuse, most of it misogynist, and which was perhaps best typified in the Gamergate phenomenon. Gavin Mueller, writing on Real Life ('No alternative'), makes an excellent case for the value of that early period of online engagement and its ability to disrupt and 'culture jam' the mainstream, but he is wise enough to note another major shortcoming: in defining itself as alternative, its value was in opposition, not in creating something new. 'Rather than articulate a clear vision of, say, a radically reorganized society, alternative [media] is cursed to indicate the locus of rebellion but without anything programmatic following behind.' As soon as the ground shifts, he says, its reason for being vanishes. 'What it leaves behind is, often enough, little more than a sneer, a prank.'

The huge irony here is that fake news itself is a species of the devil-may-care, transparently partisan engagement and alternative approach that was central to the new media period. Back then, it was largely executed by progressives and other leftists, who saw the new media moment as a chance to challenge the dominance of corporate media. Now it's neo-Nazis and other 'alt-right' practitioners spreading hate and abuse and revelling in their status as outsiders, 'the new punk', as they like to say. Gavin Mueller notes, 'the internet, which not so long ago was viewed as having helped elect the US's first black president and spread "democratic revolutions" across the world, is now blamed for stoking the flames of dangerous right wing populism'. Indeed, as the early indictments by Special Counsel Robert Mueller make clear, agents of the Russian government have also been involved, using the tools of fusion media to play upon the divisions within US politics to influence the electoral process, by creating Facebook pages to target particular US voters. To the extent that new media sought to cure the dominance of mainstream, corporate media, its techniques have now, arguably, become part of the disease.

This has all heightened the crisis of political journalism, with endless attempts to find the secret sauce that will turn political journalism into the fourth estate of our dreams. Most attempts to 'fix' the media are really elite appeals for the profession to better live up to its own, often-lauded role as a political watchdog. Sometimes these appeals are for journalists to be more open in the way they present their stories, to better engage with their audience or welcome more alternative voices. There are endless suggestions for journalists to abandon the sensationalist model of reporting I described above; and there are ongoing debates about whether objectivity should be a core value of journalism, the argument being that it is better to be transparent about your biases than to pretend they don't exist. Jay Rosen, professor of journalism at New York University, argues convincingly for journalists to 'show their work'. He says – and I couldn't agree more – that journalists can no longer pretend to have the only or the best take on a given story, that there is no perfectly objective vantage from which they report back to the rest of us. Instead, they should be transparent about the process of reporting itself, and Rosen makes 11 suggestions for how they might do this, from spelling out what they *don't* know about a story as clearly as what they do know, to being upfront about the financial cost of producing stories.

These appeals largely come from within the profession itself or from academics who are often former journalists. Some really excellent work has been done by Rosen himself, and by the likes of Emily Bell, Jeff Jarvis, John Robinson, Katharine Murphy, Jim Parker (aka Mr Denmore) and others, and they have been vital in educating – through the platforms of fusion media – the engaged audience of news consumers who are just as keen to see journalism be its best self. On any given day you can find comments on Twitter, posts on Facebook or articles on Medium calling for journalists to understand the shortcomings of everything from 'access' journalism to the use of anonymous sources, to the dangers of unexamined prejudices when writing stories, and particularly headlines. Freelance journalist Jane Gilmore's 'Fixed It' series – where she rewrites newspaper headlines to better reflect male agency in matters of sexual assault and domestic violence – is a perfect example of how structural shortcomings of the mainstream are being highlighted and challenged. Large sections of the audience now understand the way in which news is produced, and they can speak with authority on issues of narrative construction and framing. This level of knowledge and understanding and critique is fantastic, and vital, but my concern is that such suggestions will only work at the margins. The changes required cannot simply be about refining the craft; they have to be root-and-branch.

In saying that, I stress that new media and fusion media are superior to old media. However haphazardly, they represent the sort of structural transformation political journalists needed to break out of the cosy duopoly they had formed with the politicians they reported on, Trump notwithstanding. New and fusion media have done this by introducing new voices into the mix, and transforming the role of the audience in how news is gathered, distributed and understood. They have allowed individuals, groups, and organisations to intervene in the political process, and exercise some power there in a way that we need to applaud. Issues around domestic violence and the #metoo movement are prime examples of stories that would never have made it onto the front pages (if I can indulge in a little anachronism) without the ability of activists to use the tools of fusion media to put them there. Until now, the likes of Harvey Weinstein have been able to bludgeon traditional media – and his victims – into relative silence about the accusations made against him, but fusion media takes that power away from him and away from the editors who enabled the silence.

Politics itself, especially campaigning, is increasingly carried out in the interstices of fusion media spaces. Political parties – old and new, large and small – and individual politicians – from the mainstream to the maverick to the crazy – use the platforms to reach supporters and potential supporters, and a lot of this happens with little regard to the traditional mainstream media. It breeds a new type of activist and meta-political player, exemplified by the student movement that arose to fight for gun control in the wake of the Parkland, Florida, school shooting. They fulfil the prediction journalist Greg Jericho makes in his book *The Fifth Estate*, that this

> new generation will have lived their social life on the winds of the centrifugal force of social media – they will be adept at filtering the Internet's 'gems' from the noise ... The politicians and political parties who will

> triumph in the social-media world will not be the ones who seek to drive people to the centre, but rather who are able to use the forces of the widening gyre.

This is the flipside to claims that fusion media only exposes us to views that reflect our own. The risks of echo chambers, filter bubbles or even fake news are far outweighed by the positive effects of overcoming the mainstream monopoly of opinion formation and the way in which fusion media empowers citizens to better participate. Indeed, the horror stories about all the things that go wrong with fusion media are often generated by the mainstream itself and merely reflect their concerns about their own loss of power. We also, as societies, tend to panic when new technologies upset the status quo, and there has been a degree of overreaction to any new technology at least since Socrates warned his fellow citizens about the risks of the written word. Writing, he said (very definitely *said*),

> will create forgetfulness in the learners' souls, because they will not use their memories … they will be hearers of many things and will have learned nothing; they will appear to be omniscient and will generally know nothing; they will be tiresome company, having the show of wisdom without the reality.

(May the gods keep him away from Snapchat.)

And yet, for all this change, positive and negative, and for all the existential threats the mainstream media have faced, people have swung back to that mainstream as they try to navigate a politics that they find increasingly hostile to the interests of ordinary people. Gavin Mueller criticises this as a retreat from the desire to challenge corporate power. What we have seen, he says, is 'a complete reversal of the 1990s sensibilities that yearned to leave stodgy Big Media verities behind and valued a subversive and jocular underground media'. The rise of Trump, he says, has illuminated this reversal because

> intellectuals and politicos have not enjoined us to create a 'counterproject' media sphere to combat hegemonic ideology … Instead, we've been told to bolster capitalist media: to subscribe to the *New York Times*, to dutifully consume advertisements by whitelisting our favorite sites, to obtain our music from commercial platforms – so the artists get paid, of course.

Ultimately, he argues, we have re-embraced the 'neoliberal technocracy and its officially sanctioned ideologies'.

He's right to be concerned about this, but I do think his analysis overlooks the key issue of power; namely, that to stand up to power you need to be powerful, and part of the logic of the fourth estate is that it is an institution powerful enough to stand up to other sources of power. Culture jamming, blogging, alternative media all need to be cultivated, but they are not enough. Even WikiLeaks – as big and successful an exemplar of new media as you could find – which managed to gather information hugely embarrassing to power, to governments, finally had to partner with mainstream media to make an impact. It might have had the raw information, the images showing US troops firing on Iraqi civilians and the rest of it, but without the reach and institutional authority of the mainstream, its revelations had limited impact. It was only when it partnered with *The Guardian, The New York Times, Der Spiegel* and others that the story really afflicted the comfortable. The same is true of journalists who worked together to expose the scandal of offshore banking in 2016. The Panama Papers were compiled by the International Consortium of Investigative Journalists, a group of independent international journalists based in Washington DC. They put together the story, but they partnered with mainstream outlets in order to break it.

However problematic they might be, the mainstream media remain a vital ingredient in the ecosystem of fusion media. The drift of audiences back to the mainstream in the era of Trump, the call from progressive 'intellectuals and politicos' to support them, is not, as Mueller argues, 'a desire for a narrower world where corporations promise to, once again, produce a stable sense of shared reality through mass culture'. Instead, it is a cry from citizens for an institution powerful enough to represent their interests in a politics that increasingly works against them. To subscribe to the *New York Times* or donate to *The Guardian* is not to abandon the independent and anti-corporate practices of fusion media, but to assert a control over mainstream media that wasn't possible before. The more those mainstream companies rely on the direct injection of funds from their audience – via subscription and membership – the more likely they are to represent the will of those people than when they were funded by advertisers. That doesn't mean journalists will always give the audience what they want (and nor should it); but it does mean talking to them about whatever decision is made. The shift from an advertising model to a subscription model is a major structural reform and one that works to the benefit of audiences, and citizens more generally. It is this sort of media reform that we need to concentrate on.

I suspect that even improvements to journalistic craft, as important as they are, will not arise because of the compelling arguments made by journalists and media academics. They will flow from other structural changes, particularly if (when!) newsrooms become more diverse and better reflect the makeup of the communities they report on. A newsroom full of people of colour, of women, of people from a range of class backgrounds, is far less likely to fall into the traps of sensationalist reporting, false balance and phoney objectivity than one dominated by middle-class white men. Amid such diversity, it is much harder, for instance, to avoid the contradiction, encouraged by current practice, in reporting on racist acts without criticising such behaviour for fear that this would violate the norms of objectivity. We don't need to posit improvements to the craft and appeal to the better nature of journalists and editors and their role as a fourth estate so much as we need to increase the diversity of who becomes a journalist in the first place.

As subscribers, this is the sort of change we should be insisting on, and we can use the alternative space of fusion media to do it. Part of what a subscription or a membership should buy you is a seat at the table. Let's use the tools of fusion media to gather, vote for and empower a representative set of subscribers who are then installed as a regular part of the newsroom, who help oversee the editorial process, from choosing the stories to framing and presenting them. This is the only sort of transparency that is likely to make a difference: when the audience is actually inside the gates and gets to watch the sausage being made. These overseers can use the platforms of social media to report back to the rest of us and thus engender a much more meaningful and insightful discussion about what works and what doesn't. We have to stop politely asking for journalists to 'show us their workings' and instead get inside the belly of the beast so we can see them for ourselves. This is the only way we can achieve lasting, meaningful change.

It should also be obvious that we need to develop new platforms, ones that can seriously challenge the networking power of those that already exist, particularly Facebook and Google. The trouble is, the power of such platforms arises from their sheer size, and once they reach a critical mass, it is almost impossible to challenge them from below. As an article in *The Economist* ('How to tame the tech titans') points out, increasingly, the platforms 'are the market itself, providing the infrastructure (or "platforms") for much of the digital economy'. Even where new platforms have some success, such startups remain vulnerable to takeover, as happened when Facebook bought first Instagram and then WhatsApp. And the truly worrying aspect of all this is that the titans' advantage – and therefore their power – is only going to increase as artificial intelligence (AI) is loosed on the commons. The power of these platforms lies in their ability to extract and assemble data, and AI is the way they do this. It is no coincidence that Google, Facebook, Amazon and Baidu (the Chinese mega-platform) are also huge players in the field of AI. Writing in *The Guardian,* Nick Srnicek, a lecturer in digital economy at King's College London, notes, 'All the dynamics of platforms are amplified once AI enters the equation: the insatiable appetite for data, and the winner-takes-all momentum of network effects. And there is a virtuous cycle here: more data means better machine learning, which means better services and more users, which means more data.' We need to break the stranglehold these private firms have on what should be the commons of the internet. Their near-monopoly control of this vital resource must be challenged by governments, as previous technological monopolies have been.

One solution might be break them up into smaller companies, as was done in the United States with Standard Oil in 1911 and with American Telephone and Telegraph (AT&T) in 1913, which was followed with further anti-trust action in 1982 when AT&T divested itself of the Bell System company, leading to the creation of a number of smaller regional companies, the so-called baby Bells. The trouble with this approach lies in the network effects already mentioned, that platform viability and usefulness relies on the benefits of scale and that these would be lost – to consumers as well as to the companies – if they were broken into smaller companies.

Because of this, a better – if more difficult – solution might be straight nationalisation. This would keep the platforms intact, with no reduction in their network effects, but bring them under public control. Nick Srnicek argues this, as does Richard (RJ) Eskow, a writer and policy analyst. Writing in *Salon* ('Let's nationalize Amazon and Google'), he suggests that not only have the platforms assumed the characteristics of a public utility, but that they were largely built on the back of public research and innovation. 'No matter how they spin it', he writes,

> these corporations were not created in garages or by inventive entrepreneurs. The core technology behind them is the Internet, a publicly funded platform for which they pay no users' fee. In fact, they do

> everything they can to avoid paying their taxes. Big Tech's use of public technology means that it operates in a technological 'commons', which they are using solely for its own gain, without regard for the public interest.

The Economist suggests regulation alone might be sufficient to bring the platforms into line and that governments could 'make better use of existing competition law', but this strikes me as pretty unconvincing. These companies have become so powerful that they are quite easily resisting regulation, at least in the United States. In Europe, where the European Union *has* introduced some controls (the General Data Protection Regulation or GDPR), the overall effect is likely to entrench the power of the platforms because compliance costs will deter new players from entering the market. Breakup or nationalisation seems the only likely way to overcome these problems, and of the two I vote for nationalisation.

A third option uses elements of the other two approaches. Rather than target the platforms per se, the idea would be to target the data that is their lifeblood. *The Economist* writes:

> A central insight, one increasingly discussed among economists and regulators, is that personal data are the currency in which customers actually buy services. Through that prism, the tech titans receive valuable information – on their users' behaviour, friends and purchasing habits – in return for their products. Just as America drew up sophisticated rules about intellectual property in the 19th century, so it needs a new set of

> laws to govern the ownership and exchange of data, with the aim of giving solid rights to individuals.

In other words, we need to start charging the platforms for the data that they extract from us for free and without which they could not exist. There are several ways we could do this, one being the system of micropayments put forward by Jaron Lanier in his book *Who Owns the Future?* Lanier proposes a new form of property rights. 'Information is people in disguise', he writes, 'and people ought to be paid for value they contribute that can be sent or stored on a digital network.' His system would recognise the provenance of any data that we release online, either that which was created as 'a side effect of what you do to have fun online ... the videos you choose to watch' or the games you play, or through instances where you 'deliberately create data, as when you blog or tweet'. Once that provenance is created and recognised, anyone who used your data would be obliged to pay you.

Ultimately, though, I think there is an easier way to do this. Rather than forcing individuals into a commercial relationship with every platform they use, why not allow the government to deal with the platforms on our behalf and redistribute the wealth that is collected? Instead of outright nationalisation, the platform companies would be obliged to deed ownership of a certain percentage of their shares to the government, and the government would distribute the revenue earned from these shares to the rest of us. That way we all share in the wealth that we collectively create, and we are moving to the point where the platforms are functioning as public utilities. It would be a much more elegant solution than micropayments. We do a similar thing now with mining companies asked to pay royalties on the shared resources they exploit such as oil and minerals. Data is just another shared resource.

From the perspective of the news media, this would break down the almost monopolistic control that the platforms – Google and Facebook – have over the generation of online revenues, and so it would become feasible that governments could treat news as a public good in the same way that they treat health or education. They could establish publicly funded media outlets with the revenues generated from their part-ownership of the platforms. This would not simply involve increasing the funding to existing public broadcasters such as the BBC, ABC or CBC and NPR, but it could be used to fund media startups. This approach is justifiable on purely democratic grounds. The platforms have become so powerful that they are, in the words of Kate Klonick, writing in the *Harvard Law Review*, 'systems of governance' in their own right. 'These platforms are now responsible for shaping and allowing participation in our new digital and democratic culture', she notes and 'yet they have little direct accountability to their users.' The platforms famously deny that they are publishers in the traditional sense and therefore sidestep the usual norms of democratic responsibility for the speech they enable. By moving them into full or partial public ownership, we are able to reassert these norms.

One other approach worth considering as a way of pursuing structural change in the media industry is decentralisation. As I have noted, part of the established platforms' advantage arises from the fact that once they reach a certain scale – a certain number of users – it becomes very hard to compete with them, because the flow-on benefits of that scale are almost impossible to replicate: it is hard to attract a similar number of users to a new platform; it is hard to get people to switch platforms because all their friends are on the old platform; and, most importantly, people are locked into the underlying architecture of the mature platform. This means fusion media has been a period of centralisation, where the major platform companies, Google, Apple, Facebook and Amazon (GAFA, as they are often called), have effectively become closed systems. Third parties are forced to build on top of the platforms these companies provide, a situation illustrated by media companies being forced to adapt their content to the Facebook algorithm. But there is a way in which an alternative web architecture could arise, or be built, that would allow new players to compete, or provide alternative platforms for media companies to use, and it involves the decentralising capabilities of blockchain technology.

Blockchain, the tech that underlies, most famously, Bitcoin, is a set of software protocols built on top of the existing internet. Instead of being controlled by a central authority – a web platform or even, say, a bank – data is encoded in a series of 'blocks' that are then 'chained' together across a massive range of decentralised computers, so that no one person, company or authority can control it. Its integrity is maintained by this decentralisation, as it means no-one can interfere with the whole – the total of the encoded data – without changing every block in the chain, something that is almost technically impossible. This approach also makes possible the development of alternative platforms for the presentation of content such as news. So how might this work?

Blockchain is built on open-source software, which is

not only freely available but by its nature builds a community of developers and users who can have a say in how applications are developed. Participants can also earn cryptocurrency coins or tokens for their work on the network. In an article on Medium ('Why decentralization matters'), entrepreneur Chris Dixon argues that these characteristics suggest a way in which the centralised platforms of GAFA might be challenged: 'cryptonetworks align network participants to work together toward a common goal – the growth of the network and the appreciation of the token'. He argues that open systems like blockchain ultimately outmanoeuvre closed systems like GAFA because

> unaccountable groups of employees at large platforms decide how information gets ranked and filtered, which users get promoted and which get banned, and other important governance decisions. In cryptonetworks, these decisions are made by the community, using open and transparent mechanisms. As we know from the offline world, democratic systems aren't perfect, but they are a lot better than the alternatives.

Whether media organisations are actually capable of adapting to blockchain platforms – let alone helping build them – is something we can't be overly confident about. (If they aren't, they are falling into the same trap they fell into when news was first digitised and they failed to react.) Nonetheless, these options suggest that there are ways to challenge the dominance of the major platforms, and so the tyranny of the Facebook algorithm, which so besets media publishers at the moment, may only be a passing stage rather than an end point. The fact that a blockchain network has built into it the sort of decentralised, user-friendly, cooperative and transparent attributes that I am suggesting are integral for any media organisation operating in the world of fusion media is an added advantage, though we should also note that blockchain has its own inherent problems, including issues around its ability to scale, which are related to the need to verify through cryptography, thus slowing down processing. Perhaps the biggest drawback at this stage is the amount of energy needed to run them.

Despite the drawbacks, a number of startups are investigating the possibilities. The most significant of these is Civil, a blockchain-enabled platform that allows independent media sites (called 'newsrooms') to operate on the platform. Civil will be, in part, funded by the use of its own cryptocurrency, the CVL, though the individual newsrooms will be free to develop their own business models. The site intends to use the capabilities of blockchain – its publicly available registers that are difficult, if not impossible, to tamper with – to enable newsrooms and their audience to interact with each other in ways that engender trust and to avoid censorship from governments. Civil already has five newsrooms signed up, and it has to be said that they include some well-established and highly regarded journalists among their number.

Civil and other such projects are worthwhile experiments, but unlikely to be the only approach to addressing the problems that journalism faces. The ultimate solution will mean supplanting fusion media itself. Central to this change is not just the development of blockchain-enabled platforms: we need to move beyond traditional corporate and government-funded models of journalism and develop journalism cooperatives, or media organisations owned and operated by journalists and audiences that dedicate themselves to journalism's civic function.

As I've set out above, reinventing the social and organisational structures in which journalism is embedded includes:

- putting subscriber audiences in newsrooms
- diminishing the power of platforms by breakups or nationalisation
- extracting value from the data we currently create for free
- using decentralisation and journalism cooperatives to build new platforms.

These are likely to be the only sorts of changes that will alter the way the media operate and move them towards genuinely fulfilling their civic role. Paradoxically, it might also be useful to abandon the notion of journalism as 'a fourth estate' altogether. While this self-image is vital in defining journalism's civic function, it is also the cause of journalism's adversarial nature, its reliance on 'gotchas' and sensationalism, even if leavened by practices of balance and objectivity. If, instead of seeing themselves as crusaders against power, journalists saw themselves as enablers of understanding, as professionals who tend a complex ecosystem of information and relationships, a different journalism may emerge. This will only happen when audiences are central to journalism itself, and not merely observers outside the process that creates it.

But here's the final thing: fusion media are not only still embedded in the practices and preoccupations of *old* media, they are embedded in the practices and preoccupations of old *politics*. Until we change the way that politics is practised, we are never really going to change the way it is reported.

GOVERNMENT

> The only sound basis for trust is for people to have the solid experience of being served by their institutions in a way that builds a society that is more just and more loving, and with greater creative opportunities for all of its people.
>
> *Robert Greenleaf*

I don't think I will provoke an argument if I say that in most developed nations there is a palpable dissatisfaction with how we are governed. I might provoke an argument if I say that the core of this problem is voting. But that is what I will argue, so let me state it plainly: the original sin of contemporary democratic government is voting itself, and many of our current problems can be traced to this method of choosing our politicians. If we want to fix the way our governments work, the first thing we should do is replace voting with sortition in at least some of our governing bodies. Sortition means to choose – to 'sort' – by the use of lots; that is, by random sample, like the method we use to choose jurors for a court case. Instead of voting for members of parliament or congress, we should choose at least some of them randomly. It is the most straightforward way of enabling ordinary citizens to participate in the running of their country, and the effect it would have on politics and government would be transformative.

Most of us think of voting as the cornerstone of a true democracy. When a new country in the developing world moves towards democracy, we tend to judge its initial success by how soon it is able to hold 'free and fair' elections. We rejoice in this coming of age. Indeed, the Universal Declaration of Human Rights presents voting as one of our fundamental rights: 'The will of the people shall be the basis of the authority of government; this will shall be expressed in periodic and genuine elections which shall be by universal and equal suffrage and shall be held by secret vote or by equivalent free voting procedures.' But this is the whole problem: voting has come to actively undermine 'the will of the people' and we need a system that will restore their primacy. Sortition is that system. I know tampering with what we see as a basic democratic right in any way is a challenging idea to engage with, but I think we have reached the point where we have to. So let's begin by trying to understand why it is we vote in the first place.

Our fixation on voting is nothing more than a particular historical prejudice, and it is one we need to move past. Voting as a way of choosing politicians – at least, as the *only* way of choosing politicians – dates to the 18th century, the time of the American and French revolutions, and there is little doubt that the leaders of these revolutions chose voting precisely as a means of exerting elite control over the political process. Indeed, until relatively recently, only an elite of land-owning white men were allowed to vote. Even as we have fought to remove those sorts of restictions, representative government, which has become the norm in most modern democracies, has degenerated into a way for elites to maintain control over the 'democratic' process, because it is the elites – or those willing to represent the interests of elites – who are most likely to have the time and resources to ensure they are elected. If you doubt this, consider how you would fare if you decided to run for office at the next election.

French academic Bernard Manin is blunt in his book *The Principles of Representative Government*: 'Contemporary democratic governments have evolved from a political system that was conceived by its founders as opposed to democracy.' He quotes US founding father James Madison, who was equally blunt about the need for elite rule and for voting in a representative form of government (a republic) as the way to achieve it.

> The effect of representation, he observed, is 'to refine and enlarge the public views by passing them through the medium of a chosen body of citizens, whose wisdom may best discern the true interest of their country and whose patriotism and love of justice will be least likely to sacrifice it to temporary or partial considerations.' 'Under such a regulation,' he went on, 'it may well happen that the public voice, pronounced by the representatives of the people, will be more consonant to the public good than if pronounced by the people themselves …'

Seen in this light, the modern tendency of American elites to restrict the voting rights of various people through rigging electoral boundaries, demanding identification from certain citizens and restricting participation as much as possible is less an aberration than a practice with a long history.

Australia has done better than most in ensuring that voting is genuinely democratic, rather than just a means for the elite to rule, through innovations like the secret ballot and by being among the first nations in the world to extend the franchise to men without property and then to women. We also specifically decided that elections would be held on Saturdays to ensure that working people could more easily participate, a development still lacking in many countries, including the United States and Britain. Australia was one of the first countries to pay politicians a living wage so that politics was a viable job for people who couldn't rely on other sources of wealth. As well, in 1912, Australia made enrolling to vote compulsory, which may seem like an elite imposition, but it has had the net effect of ensuring that nearly everyone votes (typically turnout is around 95 per cent), which means it is much harder for Australian politicians to ignore minority voices. Australia also has a national organisation – the Australian Electoral Commission (AEC) – that controls the running of elections. It is obliged to not only make sure everyone registers to vote, but also to ensure electoral boundaries remain broadly fair. But even in Australia, with all these extra bells and whistles on the election process, the system of representative government is under stress and, along with the traditional parties, is increasingly seen as dysfunctional. At the very least, it has gone through a period of instability unprecedented since the Second World War. This has been most obvious in the way the leadership of the two major parties has changed rapidly between 2007 and 2015. In that time, a Labor prime minister, Kevin Rudd, was deposed by his own party and replaced by his deputy, Julia Gillard. Gillard herself was deposed by the party three years later and Rudd reinstalled. Prime Minister Tony Abbott, elected in 2013, was deposed by the Liberal Party in 2015 by Malcolm Turnbull. Turnbull survived the next election (with a single-seat majority) but has had to deal with constant undermining of his position by a faction within his party, led by Tony Abbott. At the state level, the NSW Labor Party went through four leaders between 2005 and 2011, while the Liberal Party has had five between 2005 and 2017. As well, Cory Bernardi, a federal senator, left the Liberal Party in 2017 and formed his own Australian Conservatives party. The leadership instability and the defection of Bernardi reflect deep divisions within the parties. And whereas once both major parties could count on receiving in excess of 40 per cent of the primary vote each during elections, both now receive approximately 30 per cent, the remaining 30 per cent going to independents and minor parties.

We have to get past the idea that voting equals democracy. It doesn't. Sortition preceded voting, and the Ancient Greeks understood precisely that it helped define the democracy they invented. Sure, the Greeks had their own problems with democracy, happily using slaves and not considering women to be citizens, but within those unfortunate constraints, they understood that if you wanted to make government truly democratic, so that every citizen's views were heard and interests considered, then you had to use sortition. As Aristotle said, 'It is accepted as democratic when public offices are allocated by lot; and as oligarchic when they are filled by election.' David Van Reybrouck explains in *Against Elections* that all legislative and executive roles in Athens were filled by sortition, as were 600 of the 700 magistrates.

To help illustrate why the founders of democracy preferred sortition to voting, consider this thought experiment. Consider an apartment building with a body corporate or homeowners association whose membership is made up of those who own apartments in the building. The annual meeting of this organisation will decide who will fill the, let's say, 20 office-holder positions available on the board. A hundred residents show up and all are asked to nominate for the 20 offices. There will be those who want to hold office and those who don't. Among those who do, some will be extremely confident and thrust themselves forward. Others will be keen but lack the gene for self-promotion, though maybe a few of them will pluck up the courage to nominate. Nominees will be given the opportunity to explain why they should be elected. 'I am a lawyer and used to this sort of organisation,' says one. 'I have lived in the building longer than anyone and know it well,' says another. 'I was on the body corporate of the previous building I lived in and have good references from those I worked with.' 'I am retired and will have more time to devote to duties here,' says another. Yet another says, 'I'm not retired, but I am rich and manage my investments from home, so not only am I available, I know how the world works.' The possibilities are endless, but you can almost guarantee that those who *present* best, seem the most self-assured or appear to have the most relevant experience will be elected.

Now imagine the same scenario, but the office holders are chosen by sortition rather than voting. Everyone puts their name in a hat and the first 20 drawn out win. This does not guarantee that the best people will end up as office holders, or that this body corporate will be superior to the elected one, but there is a completely different dynamic at work, which is likely to lead to a very different sort of body corporate. In the system based on voting, the same self-thrusting types who dominated that process are likely to dominate the business of the body corporate, and it is easy to imagine meetings tending to be arguments between strong personalities. In the one decided by lot, the whole process is likely to be much more tentative and cooperative. Any self-thrusters who make the board will no doubt still assert themselves, but their dominance will not have been normalised by a voting process that they 'won'. Given that everyone is there by luck of the draw, and that they are all eligible based on their ownership of an apartment in the building, the method of selection reinforces a sense of equality.

The point is not that choosing by sortition guarantees the best outcome, leading to the most efficient body corporate, but that it is likely to generate a much more equal and collaborative body. And that is what we want: a deliberative body of equals, even if it comes at the price of a certain level of inefficiency.

The first question that arises when we transfer this process to the level of national or state government is: are we, as ordinary citizens, really up to it? Let's look at some evidence.

It is a common belief among the elites of most democracies, those who have and are used to wielding power in various ways, that ordinary people, the lay citizens of democracy, are either disengaged from and apathetic about politics, or are so ignorant of how it works that they should not be let anywhere near the levers of power. Many elites believe we are both disengaged *and* ignorant. From the time of Plato's Republic, they have worried about 'mob rule' and the rise of a polity so governed by their base desires that good government is impossible. The modern equivalent of this ancient anxiety is often aimed at social media, which is held up as evidence of the unruliness of the democratic mob and acts as a proxy for elite concerns about the participation of ordinary voters. In outlets like *The Australian* (a News Corporation title), discussions on Twitter are painted as undermining not just democratic values but journalism itself. A November 2013 editorial, for instance, refers to the 'juvenile news-sphere, powered by Twitter' and rails against this 'mad plunge into social media-driven journalism' which, it says, 'would be mildly diverting if it wasn't so dangerous to the future of news reporting' and therefore governance.

Others express the view that the problem is 'too much' democracy. In an interview with Łukasz Pawłowski, US journalist Fareed Zakaria says, 'I think we have had too much democracy, by which I mean we've had too much of democratic procedures and too little of the "inner stuffing" of democracy; that is the liberal tradition, the tradition of protecting individual liberties against a simple will of majority.' Conservative journalist Andrew Sullivan, in a piece for *New York Magazine* ('Democracies end when they are too democratic'), quotes Plato's concerns about mob rule, while praising the 'large, hefty barriers' created by the American founding fathers 'between the popular will and the exercise of power'. He writes with approval of the various ways in which the popular will was constrained. 'Voting rights were tightly circumscribed', he notes, and that for 'a very long time, only the elites of the political parties came to select their candidates at their quadrennial conventions'. He worries that, 'Over the centuries ... many of these undemocratic rules have been weakened or abolished.'

Of course, this sort of contempt is a self-fulfilling prophecy in that the institutions of democratic participation are constructed in such a way as to exclude citizen participation – or make it extremely difficult or uncomfortable – and this lack of involvement is then offered as evidence of citizens' lack of interest in politics. The same happens with public debate on contentious issues. Excluded from mainstream institutions and from any sort of hands-on control of government processes – other than being able to vote intermittently, or to have their opinions aggregated into soundbites by polling companies – citizens use the new platforms of social media to express their frustrations. This relatively unregulated form of public speech can certainly degenerate into unedifying exchanges, but these are then taken as conclusive evidence of the unsuitability of ordinary people to participate at all. So having accused us of being disengaged, these elites hold up our actual engagement via social media as an excuse for exclusion. It is as if I called you useless for not being able to drive a car without ever giving you the opportunity to learn how.

So the issue isn't 'too much democracy' but that the institutions of the state do not properly reflect the concerns of the people, and it is this lack of input – not the inherent unruliness of the mob – that can cause democracies to spiral out of control, for demagogues to arise, and for populism to be stoked. The checks and balances that Andrew Sullivan praises the American founding fathers for installing were designed to restrain the power of individual tyrants, while the restrictions on voting and participation that he admires helped create a technocratic elite that has become self-interested and self-perpetuating as it loses contact with those it is meant to govern. Sullivan was writing before Trump was made president, but the wrongheadedness of his view is amply illustrated when he says that the 'presidency is now effectively elected through popular vote, with the Electoral College almost always reflecting the national democratic will'. In fact, Trump failed to win the popular vote and was installed by the party-selected elite members of the Electoral College. In other words, Trump was actually enabled by the very constraint (the Electoral College) that Sullivan offers as the barrier to a populist like Trump gaining power. This makes his analysis incoherent, and it points to the weakness of so many arguments about 'the mob' having too much power: 'the mob', or the popular vote, would've saved us from Trump. Rather than blame the mob and 'too much democracy', as Sullivan does, it makes more sense to blame Trump's appeal on a system of government that was seen to favour the needs of the elite to the detriment of everyone else and to which he was seen as a solution (and in saying that, I am not playing down the issues of race and gender in the rise of Trump).

Too often our democracy becomes a shouting match, an ideological shitfight in which it is increasingly impossible to tackle serious problems. This is not because of too much democracy, or because the mob has become too powerful, or because of Twitter or Facebook, but because the incentives within the current political system tend to reward – or at least, do not punish – this sort of mindless behaviour. Gadfly Australian intellectual Nicholas Gruen suggests the problem lies partly with the media, who reduce everything to 'gotchas' and 'zingers' that leave politicians too scared to offer good policy for fear of being shot down. He says that rather than argue your case, 'you're best advised to obfuscate with bland talking points designed to reassure and deaden the conversation. Or you can join the party with your own self-interested, self-righteous hyperbole. It's much more entertaining than deliberation.' Politics has become a market (there's that neoliberalism creeping in again) where politicians constantly compete for votes and the media compete for clicks or readers. But, Gruen says,

> the metaphor of the market drains the ethical content from our politics. Markets are places in which, beyond the enforcement of basic ethical norms against cheating, the whole point is to serve your own interests. I don't want to suggest that voters shouldn't consider their own interests. But democratic politics dies when it ceases also to be about how we realise the good life together.

One of the main reasons we are losing faith in our government's ability to solve major problems is because our political system is designed to exclude ordinary people. Yes, we get to vote for members of parliament and we thus get a say in who governs us. But individual politicians are largely under the control of political parties, and political parties have their own agendas which are, in turn, under the influence of other players, particularly the rich and powerful. Once elected, politicians pay lip service to reflecting the will of the people, but we-the-people rarely feel that they are really doing this. There is something pat and predetermined in how most politicians will respond to the matters that come before them that makes the whole process seem, from a citizen's point of view, farcical. Writing in *The Conversation* ('The proposed Senate voting change will hurt Australian democracy'), political scientist John Dryzek hits the nail on the head: 'Australia's federal parliament is today ... not a deliberative assembly [but] rather a theatre of expression where politicians from different sides talk past each other in mostly ritual performance. Party politicians do not listen, do not reflect and do not change their minds.' The same complaint can be made about parliaments and congresses throughout the world (and no wonder people will warm to any politician who, however briefly or imperfectly, breaks out of this ritualised performance). Gruen puts it this way: 'The ... great deficit in our democracy is not that it disregards what "we the people" think but that it is built around what we think – before we've had any time to think. The thing we're starved of isn't democratic participation, but democratic deliberation.' As Dryzek suggests, the essence of good deliberation – the chief metric of its success – is whether or not those involved are willing and able to change their minds. This is not the same thing as compromise or bipartisanship, which centrists eternally call for, both of which are just weak substitutes for true deliberation. True deliberation arises only when people come together as equals and deal openly with all the factual and emotional elements that go into making hard decisions. Party politics increasingly crowds out the ability of politicians to do this.

It can also crowd out the voice of expertise and replace it with populist appeals to tribal loyalties, and this is something else we need to guard against. In fact, we need to design institutions that make such crowding out as difficult as possible. The business of government is complex and requires all sorts of specialised knowledge, so we should be doing our best to ensure that the experts are heard and heeded. But in a democracy, a very definite division of labour needs to be observed between the expert and the lay person, between the intellectual and the citizen, between those with specialised knowledge and those who lack that knowledge, and that division of labour can be expressed very succinctly: experts are for means, citizens are for ends. In trying to work out how to achieve a particular outcome, we need to call on the help and advice of experts. But the experts shouldn't get to decide the outcome itself. Consider the issue of Australia becoming a republic, an example I'll return to below. There are many complex questions about how we would design such a system. To answer those, any sane nation is going to draw on expert opinion. But the experts themselves don't get to decide *if* Australia becomes a republic. That final decision, the goal of the deliberation, is down to the people themselves.

The link between the expert and the layperson – the key process by which the two groups may work together – is deliberation, both sides talking with each other in a vaguely rational way and coming to mutually agreed conclusions. I say 'vaguely rational' not to be sarcastic or to downplay the importance of such rationality, only to recognise that none of us, no matter how wise or, indeed, rational, is ever really driven by rationality alone. Our emotions are also involved, and we need to recognise that, not as a failing but as an integral part of the decision-making process. If we accept that, *rationally*, our individual vote makes almost no difference in the greater scheme of things, why do we bother to vote at all? In large part it is because we are motivated by our emotional investment in the process and the issues at hand. As philosopher Martha Nussbaum has said, 'All political principles, the good as well as the bad, need emotional support to ensure their stability over time.' The trouble with our current system is that it is optimised for emotion; that is, politicians and media – and all the other pressure groups that seek to influence us – don't look to persuade us with compelling logic, but to sway us with emotional appeals. What we need, then, are institutions that recognise both aspects of decision-making – the rational and the emotional – and that allow deliberation to reconcile these aspects. We again need to invoke the organising principle of the commons, reaching out beyond imposed, top-down solutions and instead allowing the meaningful play of different ideas aimed at an outcome satisfactory to all.

To that end, I propose that we create a new chamber of parliament that I'll call the People's House. What makes it different it from other houses of parliament is that its members will be chosen by lottery, or sortition. Any adult citizen could at some stage of their lives be called on to serve in the People's House in the same way that we may all be called upon to serve on a jury. Instead of deciding the outcome of a criminal trial, the members of the People's House will deliberate and vote on legislation, and therefore decide how to run the country.

In polities around the world, there is now a substantial body of work involving so-called citizens' juries, deliberative polls or other forms of elite–popular discussion. They show very clearly that a house of parliament populated by ordinary citizens chosen by sortition could work, and work well. A deliberative poll, for example, attempts to inform opinion by providing the opportunity for wide public discussion among those polled. This is done by bringing together a representative sample of the population – selected by a polling company – for two days at a central venue where they can discuss the issues with each other in small groups, as well as put questions to a panel of experts who represent a range of opinions on the topic at hand. Participants are polled before and after this two-day meeting, and the second set of results is taken to be indicative of the influence of the deliberative process. The two-day deliberation is preceded by approximately six weeks of preparation by the citizen participants. They are issued with a 'briefing document' that outlines the issues involved and the arguments for and against. They are also encouraged to follow the issue as it arises in the media and to discuss it with as many people as possible.

A deliberative poll was held in 1998 on the topic of whether Australia should become a republic. The number of people originally approached for the poll was 1220, and they were asked to participate in a 25-minute phone interview. After this interview, 770 people said they would be interested in hearing more about the process. Follow-up information was sent out, along with letters to employers to request time off work so people could attend. Ultimately, 347 people accepted. According to Issues Deliberation Australia (IDA), which ran the poll, 'Australians were jumping at the chance to be involved ... a final sample of 347 representative Australians arrived in Canberra on October 22nd, 47 more than our original goal.' This high turnout may have been attributable to the persuasive techniques used by the organisers, but it also attests to the fact that participants perceived the issue to be important and the forum to be credible. This genuine willingness to participate contradicts the strand of thinking set out earlier in the chapter suggesting citizens are apathetic and motivated only by self-interest. I was one of those who sat in the public galleries watching the two days of deliberation and I can say that it was a truly inspiring event. Ordinary citizens revelled in the chance to question the various experts gathered for their benefit, and as their confidence grew, they were quite willing to challenge the information they were being given. On occasion, the debates became quite robust, though they maintained a high level of decorum and respect. At one stage I was sitting next to journalist and commentator Bob Ellis and I asked him what he made of proceedings. 'Fucking amazing,' he said as he stood up and left the chamber, shambling out like some press gallery Pig-Pen. (Vale, Bob Ellis.)

Commentator and constitutional lawyer George Williams spoke on the revelation of citizen involvement on ABC television:

> Australians are often criticised for being apathetic about their political system. I think this event exposes that for what it is: that is, if Australians are given an entry point into the debate, they want to be engaged; they want to be involved, and the people here have shown tremendous enthusiasm and commitment to seeking out the issues.

This runs counter to so much discussion of the role of citizens in public debate, as Williams points out, that it would be remiss not to note the extent to which the deliberative poll was a learning experience for the 'intellectuals' as well as the citizens. The various experts and other elites involved had their preconceptions challenged, including the key expert participant, Malcolm Turnbull. At the time, Turnbull was the most public of the 'public intellectuals' involved in the republican debate. He contributed articles on the topic, wrote a number of books and was constantly interviewed in the media. In addition, he formed the Australian Republic Movement, in part a public organisation aimed at promoting the yes case, and also a lobby group that had the ear of all major parties. Turnbull had a great many misgivings about the deliberative process, mostly based on his fear of citizens' ignorance. The deliberative poll experience did not change Turnbull's general assessment of the public's ignorance; nonetheless, he did acknowledge the change of opinion that it recorded and recognised that once 'the delegates understood that the President would be filling the Governor-General's shoes and should therefore be non-political, they recognised that a direct election was not the right method of appointment'. This grudging admission says more about Turnbull's ingrained elitism than it does about the ability of the participants themselves, and highlights that people are more than capable of sifting through the various claims made by expert participants if given the chance to do so.

The reaction from the non-expert citizens involved was extremely positive. IDA received a large number of thank-you letters from participants, and in interviews afterwards, participants expressed consistent feelings about the value of having been involved. One participant from Victoria put it this way:

> Can I just say that as an elder citizen that I've been tremendously informed and stimulated by this gathering. I would just like to say how wonderfully

> I've seen the democratic process at work. On Friday … there were only one or two spokesmen [but] by this morning's session every one of our fifteen delegates was speaking vociferously and strikingly at times.

The empirical evidence bears out the fact that, on average, participants' general knowledge of the issue at hand and even of the general political process improved greatly, but it is the less tangible 'sense of involvement' that is also enhanced by the process. Many participants commented that they were clearly impressed – even grateful – for being included in such an event. I was studying the poll as part of my PhD, and it was extraordinary how often participants predicated their questions in the plenary sessions with an expression of gratitude for their involvement.

Such forums cannot abolish the division of intellectual labour – lay people don't suddenly become experts – but they can make discussion between experts and non-experts more equal. As such, the forum potentially does much more than improve the public's general and specific knowledge of the issue at hand: it provides a forum of cooperation and deliberation that helps engender trust and respect among participants. It tends to break down, from both sides, the tendency for experts and non-experts to view each other as adversaries – where the experts view the citizens as merely ignorant and a slate to be written upon, and where the citizens view the experts as an elite merely asserting the power that arises from superior knowledge. Those operating as experts in this environment are not simply articulating their own views in a way that a lay audience can easily understand. They are making available their knowledge for a lay audience to reach their own conclusions about the issue.

This is what the deliberative poll does most successfully – it empowers ordinary citizens – and it provides a useful model for other institutional design.

Nicholas Gruen outlined some further examples of such a model in his 2017 address at the Adelaide Festival of Ideas. Gruen is the former chair of the Australian Centre for Social Innovation, which has run a number of citizens' juries in South Australia. A citizens' jury is similar in design and function to a deliberative poll, though it doesn't put the same emphasis on polling participants before and after the process of deliberation. Gruen says citizen juries 'draw people in and, energised by the invitation to participate, they're keen to give of their best'. Participants felt lucky to be invited, with one commenting: 'All the jurors in the room had a real sense of responsibility. I don't know whether that was usual, but I was blown away by it.' Another was surprised that they found 'way more capability in the community than we give credit for'. Gruen also notes that the experts and special-interest groups involved were 'impressed by jurors' eagerness to do a good job and felt that jurors were well informed, well read, and offered thoughtful questions'. He says that many of the jurors 'increased their interest in policy issues', and that they all 'shared a desire to do it again and recommend the experience to others. This included sustaining contact with the group which became quite important for some.'

The lessons we learn from these experiences with deliberative democracy is that extending them into a more formal and permanent part of our governing process is worth thinking about seriously, and any claims that such a concept could never work because ordinary people are disengaged or apathetic should be treated with the contempt that these examples suggest they deserve. Nonetheless, we need to acknowledge examples of participation and non-elite attempts at governance that show up certain limits to these alternative approaches.

The most obvious are the movements associated with Occupy Wall Street, the protest movement that began with a handful of people in Bowling Green Park in New York City before moving to Zuccotti Park on Wall Street, and then was replicated in cities throughout the world. David Van Reybrouck explains in his book *Against Elections* that the 'general assembly lay at the heart of the movement and it quickly developed its own arsenal of rituals, most striking among them the "people's mic"':

> Because amplification was banned, everything was acoustic, without technical aids, even at meetings with many hundreds of participants. A person would speak and the people around would repeat what they said until the message reached those right at the back in a series of waves. To express agreement or disagreement or to ask for further elucidation, various hand signals were invented. The meetings had no chair, no leaders of factions, no spokespeople, at most a few moderators to keep the process on track. Horizontality was the name of the game.

This is the same 'horizontality' I criticised in the Commons chapter, and which any number of sympathetic but disappointed commentators and intellectuals – including David Harvey and Slavoj Žižek – saw as a fundamental weakness of the protest movement. Author Thomas Frank directed his scorn (found in the *Baffler* article 'To the precinct station') at the horizontal engagement and decision-making process that the protesters employed. 'How did OWS blow all the promise of its early days?' Frank asks.

> Why do even the most popular efforts of the Left come to be mired in a gluey swamp of academic talk and pointless antihierarchical posturing? … Occupy Wall Street … took its horizontality seriously. It grew explosively in the early days, as just about everyone with a beef rallied to its nonspecific standard. But after the crackdown came, there was almost nothing to show for it.

So if we are going to build a new People's House, it cannot submit to what writer Christopher Lasch once called the 'cult of participation'. Its goal should be to foster deliberation, but it must be able to translate that deliberation into action, otherwise it is just another fob to the citizenry, an empty symbol of democratic involvement.

Interest in sortition is growing, and a relatively large number of books, papers and articles have put forward ideas for the design of different deliberative bodies populated by candidates selected by lottery. My aim here, then, is not to adjudicate between them, but simply to indicate the range of possibilities available. The specific model used in a given country or jurisdiction should be a product of genuine deliberation, as should the design of the chamber itself. In thinking about this, we should note again that sortition is already how we select juries and this has proved to be a remarkably robust institution: and to the extent that it has been curtailed as a practice – with some sorts of trials, for instance, exempt from juries – this can best be viewed as elite encroachment rather than any particular weakness with the jury system per se (but that discussion is beyond our scope here). There are also the citizens' juries and other similar one-off deliberative exercises, and all of these reinforce the viability of sortition. Still, extending the idea into actual government will require new formal structures, possibly as adjuncts to our various houses of parliament, perhaps even as replacements for them.

Some suggestions for incorporating sortition into the political process are more radical than others. John Burnheim, in his book *Is Democracy Possible?* sees sortition as a way of almost replacing the state altogether, and others, such as economists Paul Cockshott and Allin Cottrell have developed and discussed his ideas. Burnheim's main goal, as is mine, is to stress the importance of creating a truly deliberative body, and he is clear that the foundation of that body needs to be sortition. He says decision-making bodies should be 'statistically representative of those affected by their decisions ... Elections ... inherently breed oligarchies. Democracy is possible only if the decision-makers are a representative sample of the people concerned.' L Léon, who uses the term 'lottocracy' to describe his preferred system, takes a similar approach. But Léon is aiming for world government and falls into the trap of deciding that what is obvious and desirable to him is therefore inevitable. He says in his online book *The World Solution for World Problems*:

> By simply applying common sense it is very well possible to develop a fair and workable system for our world government. A system that nobody can possibly object to, and that will end all politics, diplomacies, opposition parties, ministries of defence, wars, etc. and

> therefore, all secrecy, the arch-enemy of co-operative existence.

I admire the sentiment but am inherently wary of any reform that tries to pass off such reform as 'common sense' to which 'nobody can possibly object'. As if.

Among those who wish to apply sortition to already existing legislative bodies are Terry Bouricius, who used to be in the Virginia State legislature; Ernest Callenbach and Michael Phillips in their book *A Citizen Legislature*; and French activist Étienne Chouard, who outlines his proposals in an article, 'Democracy through multi-body sortition: Athenian lessons for the modern day'. Some suggestions, such as that by political scientist Robert A Dahl in his book *Democracy and Its Critics*, involve using deliberative bodies as an adjunct to existing parliaments, gatherings that could 'hold hearings, commission research, and engage in debate and discussion'. I made a similar suggestion in my doctoral thesis, submitted in 2001, arguing that deliberative polling could be incorporated as a regular feature of the committee system of the Australian Senate (which I still think is a good idea, though I now think this is an insufficient reform).

Others argue for deliberative bodies chosen by sortition to replace existing legislative chambers altogether. Perhaps the most ambitious mainstream suggestion to replace an existing chamber of government comes from Anthony Barnett, Peter Carty and Anthony Tuffin. These establishment academics propose replacing the unelected members of the British House of Lords with members chosen by sortition. Their plan was first set out in a submission to the 1999 Royal Commission on the Reform of the House of Lords in the United Kingdom and later published in a book,

The Athenian Option. They propose that the House of Lords 'should be changed into a second chamber with broadly similar powers to those exercised by the present one' and that

> the reformed second chamber needs an impartial, nonparty political character. This can be obtained by selecting a proportion – ideally, ultimately the majority – of its members by lot from among registered voters, on the lines of a jury. This should not be entirely random. Different regions should be represented in proportion to their population, each with an equal number of men and women.

My preferred approach is something along these lines in any jurisdiction with an upper house. In an Australian context, Nicholas Gruen has developed a detailed proposal. He suggests a People's House made up of '195 citizens chosen by lot for six-year, non-renewable terms with 65 appointed every two years'. The house would have the power to delay but not block legislation passed by the House of Representatives, the lower house in the Australian system, which would continue to be elected by popular vote. This power is similar to what the British House of Lords currently exercises and it can be effective.

However, Gruen believes, and I agree, that to get the full deliberative and democratic benefit from a People's House, it would ultimately need to have real legislative power. The Australian Senate, at least initially, could simply be selected by sortition and retain the powers it currently has to review and reject legislation. Gruen suggests another power for a people's chamber: 'to compel a secret ballot of the other chamber(s) on the matter in question'. If there was still a deadlock, a power similar to section 57 of the Australian constitution would be invoked, and after three months of deadlock, 'the matter would be resolved in a joint sitting between them'. That is, both houses would sit together, deliberate and vote. This would mean the lower house could not ignore the advice of the people's chamber – a huge risk if it was established merely as an advisory body. Knowing that the People's House could not only reject legislation but was likely to compel a secret ballot, and perhaps a joint sitting, the lower house has a genuine incentive to revisit the legislation and look for a solution, for minds to change.

Acknowledging that convincing our current elected representatives to create a people's chamber is going to be difficult – akin to getting turkeys to vote for Christmas – Gruen suggests an interim measure whereby citizens themselves, using donations from the public or businesses, instigate their own people's chamber. He asks us to imagine a body of 99 people chosen at random from the relevant population sitting as a citizen's chamber:

> It would be given the resources for some research capacity and to call expert witnesses. One might sit four times a year for nine days – encompassing one week and the weekend on either side of it … The chamber would deliberate on the choices before the electorate at large … They would deliberate on the parties' policies, they would invite spokespeople for those parties or others to address them and help them deliberate on the merits of the choices to be made by the people. They would also … work on brief

> communiqués explaining the case for and against various options together with how the body voted on them.
>
> I think this could have an electrifying effect on political business as usual.

I think it could too, and if we are really serious about bottom-up reform of our democratic institutions, then reforming the seat of government itself in this way, a way that installs ordinary people at the heart of power, is essential. Our neoliberal economy and the representative form of government that dominates our societies do everything they can to divide us from and pit us against each other. A People's House transcends these divisions and brings us together. The basic concept of sortition is pretty straightforward, and introducing it as a replacement for voting in, say, the Australian Senate, while leaving that body's other powers intact, represents, at least administratively, fairly minimalist change. But on every other level, the potential effect is explosive. In one fell swoop you diminish the power of the parties and that of many of the lobbyists who exist to influence their decisions. You transform the way in which the media covers politics. You hand control of at least part of the legislative process to a genuinely representative sample of the population as whole, rather than vesting it in a bunch of elites and their representatives. You empower people in a way that the current system could never hope to do, and you reconnect our chief democratic institution with the life in common.

Nothing is going to change until the main source of power in our society, our seat of government, is populated by people who are genuinely representative of the society at large. We have been taught forever that the way to do that is by voting, but that is simply wrong, and the quicker we unlearn it the better, no matter how counterintuitive it might seem at first. If you want a truly representative government of, by and for the people, then you need to choose it not by voting, but by sortition.

Now, that is power.

WEALTH

> 'I just want my fair share – which is all of it.'
>
> *Billionaire businessman Charles Koch*

In 1942 – that is, during the Second World War – the British government released a report called *Social Insurance and Allied Services*, more usually called the Beveridge report after its main author, economist William Beveridge. It is the founding document of what became known as the British welfare state. It sets out in detail the principles and practices of wholesale reform of the British economy, including the creation of the National Health Service, and it addresses itself specifically to solving what Beveridge called the 'giant evils' of squalor, ignorance, idleness and disease. The report is organised around three guiding principles: first, that reform should be comprehensive, not directed solely at sectional interests; second, that social insurance should not just attack want, but also the other giant evils; and third, that any way forward must encourage cooperation between the state and the individual. Beveridge writes: 'The State in organising security should not stifle incentive, opportunity, responsibility; in establishing a national minimum, it should leave room and encouragement for voluntary action by each individual to provide more than that minimum for himself and his family.'

It is incredible to think that the writing of such a report was possible as Britain faced the threat of destruction, and there was certainly debate in the British parliament about whether such a paper should be published, and whether the nation was able to afford its recommendations. But it made perfect sense. As I have noted, history shows that societies only generally address inequality when faced with existential threat – brought on by war, famine, disease or revolution – and indeed, Beveridge himself recognised that the moment in which he found himself was a one to be seized: 'Now, when the war is abolishing landmarks of every kind, is the opportunity for using experience in a clear field. A revolutionary moment in the world's history is a time for revolutions, not for patching.'

We can safely say that the creation of a welfare state on the scale imagined by Beveridge was no patch. It fundamentally rethought the way in which wealth was distributed in British society and was a model for other approaches taken around the world. Over the years, in Britain and elsewhere, the effectiveness of the welfare state has been whittled away and undermined, so it is again time for a revolution, not for patching. That revolution is universal basic income (UBI). It, too, addresses itself to giant evils, first and foremost the growing inequality and insecurity that is plaguing developed nations and undermining the viability of democratic society. Unlike Beveridge and most of the world in 1942, we are not in the middle of a declared war, but the question we have to ask ourselves is, do we need an actual war to motivate us to undertake the radical reform that is needed? I see embedded in the growing calls for UBI the early warning signs of a world under stress and looking for answers. People sense that violence is threatening – or becoming more common – and so UBI is a siren call. It is crying out to be the response made *instead* of war, not as a result of it.

UBI, then, needs to be seen as more than just a methodology, a means to an end. In the words of Shannon Ikebe, writing in *Jacobin*, UBI is 'a demand – a demand that exposes the irrationality of an economic system in which productivity increases seem to bring more unemployment and misery instead of the expansion of freedom they make possible'. Advocates recognise that UBI isn't simply an alternative way of distributing wealth but represents a more fundamental reorganisation of society; UBI's tendency to unsettle the way we do things now is a feature, not a bug. UBI is the revolution before the revolution, one of the ways in which we might reform our broken politics and economy without descending into chaos.

The definition of UBI offered by the Basic Income Earth Network (BIEN) – which is the most active advocacy group for UBI in the world – is straightforward, but it encapsulates a radical idea. BIEN defines it as 'a periodic cash payment unconditionally delivered to all on an individual basis, without means test or work requirement', and that is the definition I will use here. Every aspect of BIEN's definition captures something particular about what a UBI offers, which is important to understand because, as we will see, many of the criticisms of UBI end up being nothing more than straw men based on conflating UBI with other forms of welfare. So the fact that the payment is periodic is important, because the regularity provides ongoing financial stability and therefore the ability to plan. That the payment is delivered to individuals and not families or households underlines the fact that UBI recognises individual freedom.

But the three key ingredients of the definition are that it is basic, universal and unconditional. Basic means the payment is sufficient to cover the essentials of life. In Australia, the single-person pension (currently just over $21 000 per annum) might be seen as an appropriate level, though it would vary from country to country. It is universal by being equally available to the poorest member of society as it is to the startup billionaire. The payment is unconditional in that no-one has to fulfil any obligations in order to receive it, and you're still eligible if you earn other income. The idea that UBI provides a floor but not a ceiling on income is central.

Of course, defining a radical project is a lot easier than implementing it. I have sat in rooms – at conferences, workshops and roundtables – over the last few years talking about UBI with businesses, unions, bureaucrats and academics, and I have heard every manner of objection, ranging from 'we can't afford it' to 'it is politically impossible', to 'it will make people lazy', and they are (mostly) legitimate concerns. The thing is, though, all those criticisms could also be applied to just about every other form of policy that addresses itself to inequality or social security, and as such, they can't stand as reasons not to pursue UBI. If you object to UBI, the onus is on you to explain why people would be worse off if we paid them a regular basic income, and very few criticisms of UBI even attempt to do that. So let's dispense with a few of the weakest objections first.

The weakest of the weak objections to UBI is the claim that paying people an ongoing, no-strings-attached income will simply see them reject work and choose to 'do nothing'.

This is an update of the 'dole bludger' argument, the eternal fear that a significant section of society simply wants to suck on the teat of the welfare system funded by *your* taxes, and it will be hard to shake given how impervious prejudices tend to be to empirical evidence. Nonetheless, UBI trials show, remarkably consistently, that fears of people opting out of work when you pay them a basic income are unfounded. Participants in a UNICEF-run UBI trial in India in 2013, for instance, actually worked more. As Guy Standing, the professor from London University who supervised the trial writes: 'There was a shift from casual wage labour to more own-account (self-employed) farming and business activity, with less distress-driven out-migration. Women gained more than men [and there] was an unanticipated reduction in bonded labour. This has huge positive implications for local development and equity.'

Other trials around the world have borne out these results, to the extent that it can no longer be accepted as a legitimate concern, certainly not in the 'common sense' way in which it is trotted out. Ioana Marinescu, assistant professor at the University of Chicago Harris School of Public Policy, who has examined the way in which unconditional cash transfers – everything from lottery wins to UBI-type schemes – affect people's willingness to work says that 'the impact on work is very, very small'. In an interview with *Fast Company* magazine, she says that data from around the world gives 'fairly consistent results' and that this consistency is 'deeply encouraging'. Although, she says, 'In principle, we can't expect the same policy to have the same effect [everywhere]', the reality is that 'across these studies we see fairly similar effects'. And note something else about the results from the UBI trial in India: the *nature* of work in the community changed, so that rather than choosing *not* to work, recipients of basic income instead chose to be self-employed, to work on their own terms. This not only refutes the UBI-will-make-us-lazy argument, but supports the idea that UBI liberates people in a way that more conventional welfare systems don't. It suggests that while UBI may provide a safety net for anyone forced into unemployment, it can also be work-positive in that it provides enough financial security that people are willing to pursue other work options.

Perhaps aware of how weak the argument is, or perhaps because they don't want to be accused of having a low regard for their fellow citizens, critics often express the laziness argument in another way, couched as 'we don't want people to become dependent on government'. But this, too, bears no scrutiny. The current system of welfare payments, hemmed in as it is with what are generally called 'mutual obligations' – a series of tests and conditions recipients must pass or fulfil in order to receive payment – does far more to make people dependent on government than an unconditional payment like UBI does. Take, for instance, the basic requirements Australians must fulfil in order to receive unemployment benefits:

> In return for activity-tested income support payments, job seekers must meet their mutual obligation requirements, which include:
>
> - entering into a Job Plan,
> - fully complying with the terms (requirements) in their Job Plan,
> - demonstrating that they are actively looking for suitable paid work,

- accepting offers of suitable paid work,
- attending appointments with their employment services provider,
- attending all job interviews,
- attending approved education or training courses or programs designed to address any barriers that a job seeker may have to entering the workforce, and
- never leaving a job, training course, program or other required activity without a valid reason.

Job seekers with mutual obligation requirements will be subject to the job seeker compliance framework if they fail to meet their requirements.

Beyond these, there are lists of further requirements depending on whether you have a disability, whether you are a carer, whether you are a student, whether you are over 55 years of age, or even whether you are an early school leaver. The entire welfare system in most developed countries is so riddled with hoops like these, which people have to jump through to be paid what is often a pittance, that it renders them by definition not just dependent on government – you don't get paid unless you do what the government wants – but turns them into among the most surveilled citizens in the country. If anything, these regimes of compliance are increasing, and in some jurisdictions they include drug testing and cross-referencing of claimants' applications with their tax records. Indeed, the current Charities Commissioner in Australia, former Labor minister Gary Johns, a few years before taking the role as commissioner wrote a paper arguing that people who received unemployment benefits should have to take contraception as a condition of payment. If any of this massive and expensive bureaucracy results in the government making a mistake in your payments, you are entitled to complain, but if you go public with that complaint, the government reserves the right – written into legislation – to make public your personal details. This happened in 2017 to a freelance journalist who wrote an article detailing her experience with the error-bound process. The relevant department – Centrelink – then provided details of her case to another journalist, who wrote a story that exposed her personal information to public view.

You don't get to say you are concerned about people becoming dependent on government with UBI when you support (actively or by default) a system of welfare that is already chock full of 'mutual obligations'. Conditional welfare *is* dependence. It is only when you take away the conditionality that you free people from it. Philippe Van Parijs, in his book *Basic Income*, goes so far as to note that conditionality creates 'an intrinsic tendency to turn … beneficiaries into a class of permanent welfare claimants'. So, even couched in these more 'positive' terms of not wanting people to become dependent on government, the laziness argument doesn't hold water. People do not become lazy just because you provide them with some financial security, and they are actually more dependent on government when every payment they receive is conditional on government approval.

Generally, this laziness/government dependence argument comes from those on the political right, but some of the weakest criticisms of UBI actually originate on the political left. Increasingly, left-wing critics of UBI paint it as a right-wing Trojan horse, a libertarian plot hatched in Silicon Valley to dismantle the welfare state as we know it. The fact that entrepreneurs such as Mark Zuckerberg, Richard Branson, Sam Altman and Elon Musk have expressed interest in what they call a UBI is offered as proof of its undesirability. The bogeyman most often conjured to scare us away from UBI is Milton Friedman, the US economist famous for his hatred of government in general, and one of the intellectual godfathers of the Thatcher and Reagan dispensations. The rhetorical value of saying, as some on the left do, 'Do you really want to implement something that is supported by the likes of Milton Friedman?' is pretty high – full marks for scaring the horses – but it is also misleading. The form of UBI favoured by the likes of Milton Friedman, indeed, of most of these bogeymen, is not really a UBI at all; it is a negative income tax (NIT). This is 'a progressive income tax system where people earning below a certain amount receive supplemental pay from the government instead of paying taxes to the government'. It is tied to employment, and the amount a person receives tapers as their income from paid work kicks in. An NIT provides a floor below which people's income can't fall, but it's neither universal nor unconditional. If you are rejecting UBI 'because Milton Friedman likes the idea', then you aren't really rejecting UBI.

The other weak argument from the left involves a rejection of universality. Chris Bowen, at the time the Australian Labor Party's shadow treasurer, has said, 'A universal basic income would be just that, universal. Providing payments to millionaires, at a considerable cost to the taxpayer.' Again, this has some rhetorical force, but rejecting universality as a political principle is a dangerous route for anyone on the left to take. After all, the left doesn't reject the universality of government healthcare or public education, and so rejecting it for UBI is, at best, inconsistent.

Nonetheless, Australia has a highly targeted system of welfare transfers, which Bowen has raised as a point of pride. Welfare economist Peter Whiteford says, 'Overall, Australia has a relatively low level of spending on cash benefits but [it] concentrates these benefits on low income groups more than any other rich country.' This targeting helps keep overall tax rates down, and arguably leads to an efficient system of transfers. But it is not without its problems. For low earners the combination of income tax and benefit withdrawal rates also tends to create poverty traps. That is, as someone's income from paid work increases, the concomitant tax rate – the effective marginal tax rate – eats up the extra income they earn. This is why people talk about this situation as a disincentive to work: why take the job if it means you will earn less once the effective marginal tax rate is taken into account? UBI addresses this problem directly. Because the extra money you earn does not affect the payment of a basic income, there is no poverty trap. Indeed, the evidence from UBI trials suggests that the ability of people to increase income by taking a job, even a part-time one, increases the likelihood that they will, in fact, take the job. The disincentive to work is reversed.

But there are other equally compelling reasons to make basic income universal, to make it available to Bowen's millionaires, the first of which is that universality helps turn such people into supporters of the scheme. It helps cement the support of the wealthy even if, as is the case with Medicare, they buy cover in private markets. Remember, if UBI is financed through taxation, the wealthy end up being net contributors to the scheme. They put in more than they get out, but by making them recipients of the benefit – via universality – we increase the likelihood that they will support the scheme in the first place. This is exactly why Medicare and the UK's National Health Service were made universal, and it works. The second is that universality helps remove the stigma associated with 'being on welfare': when everyone is getting the benefit, no-one need feel as if they are somehow taking advantage of their fellow citizens. So just as no-one feels shame in using a public hospital, no-one will feel bad about accepting their share of a *universal* basic income. Critics who reject UBI on the grounds that it 'provides payments for millionaires' are being disingenuous.

The more meaningful objections to UBI from the left go to the heart of the structural challenge it poses to an economy. A very big concern of progressives is that a UBI would be funded by discontinuing other services – everything from healthcare to pensions – and forcing citizens to buy those services in private markets. They also worry that a cash payment encourages a shift away from collective solutions and helps entrench an individualist approach to welfare provisions. Responsibility for any shortcomings could then be blamed, more than ever, on individuals rather the structural obstacles they often have to deal with.

Such concerns are why the idea – put forward by the more gung-ho basic income advocates – of UBI being a one-stop solution to welfare design need to be challenged, and indeed, the very strong feeling of a UBI workshop I attended at the University of Melbourne in August 2017 was that any attempt to sell UBI as a silver-bullet solution is dangerous overreach. Frances Flanagan, a researcher with the union United Voice, suggested that a UBI 'can't bear too much weight as a single policy fix', and this view was echoed through much of the workshop discussion. Academic researcher Andrew Scott, for instance, stressed that UBI is not 'a panacea independent of other policies, especially in regard to housing and childcare'. Perhaps these concerns are most clearly put by former trade union leader Tim Lyons, who now works at the Per Capita think tank, in an article on its site:

> I am deeply unconvinced by the push, supported by an increasing number of people on the left, for a Universal Basic Income. It is of course, originally a right-libertarian prescription, which would allow for the state to be dismantled, as all would have an income sufficient to access what they need in a market. It is not a proposal supported on the other side on the basis that it's 'as well as' the welfare state but rather it's explicitly intended to be 'instead of'. UBI also makes no sense using a power analysis – imagine a world where we had enough power to restructure the tax system to collect the revenue (billions of dollars in additional taxes on capital and very very high marginal rates of personal income tax at the top end) to fund a UBI. Is it really the case that what you would do with the money is cut everybody a cheque?

Lyons is right to suggest that right-libertarians see such schemes as a way to dismantle various welfare services and force people to instead buy these services – health, education, housing and others – in private markets. Friedman saw his NIT scheme as a way of removing what he believed were 'wasteful' welfare programs. In his book *Free to Choose* he writes:

> The program (NIT) has two essential components: first, reform the present welfare system by replacing the ragbag of specific programs with a single comprehensive program of income supplements in cash – a negative income tax linked to the positive income tax; second, unwind Social Security while meeting present commitments and gradually requiring people to make their own arrangements for their own retirement.

His latter-day disciples at organisations like the Adam Smith Institute, a British think tank, preach the same gospel on their website: 'The economic benefits of a NIT in this day and age are thus substantial, slashing bureaucratic costs, in addition to the political and social benefits of decreasing the size of government and promoting individual liberty.'

So there are legitimate concerns that an NIT is a Trojan horse sent to dismantle the welfare state. But let's look at these objections more closely. They turn on the actual design of the UBI, which is precisely why I stressed at the start of this chapter the importance of being careful about how we define UBI. An NIT and a UBI are *not* the same thing, and it does no-one any good to conflate the two. So while it is fair to say that an NIT is a 'right-libertarian prescription', it does not follow that UBI is. Indeed, given that various forms of basic income can be found recommended in Thomas More's *Utopia*, in the writings of Thomas Paine and lauded by the likes of Martin Luther King, Noam Chomsky and Archbishop Desmond Tutu, the tag of 'right-libertarian' seems misplaced.

There is absolutely no reason a UBI needs to be offered in a form that involves the shutting down of other forms of welfare. None. Certainly, a UBI involves a major overhaul of current welfare payments and the tax system that supports it – this is an inherent and desirable aspect of the demand that it represents – but that is not the same thing as the wholesale closing down of all other welfare as envisaged by supporters of an NIT. To object to it on the grounds that UBI might lead to the loss of these other services is like objecting to income tax because some people want to set the rate at 20 per cent but you think it should be at 30 per cent. That is an argument about the design of the tax system, not about tax per se. You don't reject the very idea of 'income tax' just because someone wants to set it at a different rate. So concern about whether or not UBI will be used to replace other forms of welfare is legitimate, but it tells you precisely nothing about UBI as a concept, only about the importance of how the scheme is designed and implemented.

I accept that many of the left have genuine concerns about UBI, though I hope the preceding discussion has at least addressed – and maybe answered – some of them. At the very least, I think the left should stop trying to turn people off UBI by conjuring the ghost of Milton Friedman, or by conflating UBI with other schemes like a negative income tax. Those arguments are unworthy. Regardless, I think there is another very practical reason people on the left should consider embracing UBI as the foundational idea of a new big-picture program for reform. The risk the left runs in rejecting a UBI scheme in any shape or form is that it actually leaves the field of welfare reform open for the right to occupy unchallenged. For the truth is that a scheme like an NIT – or some other variant of UBI – has considerable appeal to 'small government' conservatives, exactly as the left warns; if it is combined with an appeal to lower taxes, there is always the risk the idea will take hold among those with influence in conservative policy circles and ultimately find its way to implementation.

Beyond this, it is very difficult to see how the left can reject outright a scheme that puts a floor under the lifetime income of citizens in general. Indeed, the Australian Council of Trade Unions' current push for a living wage is motivated by similar concerns, but a UBI offers the benefits of that idea in a more streamlined and efficient manner and makes it available to more people. Because UBI is paid without imposing conditions, a lot of bureaucratic intrusion into people's lives simply disappears, and I don't think we should underestimate the difference this can make to the lives of ordinary people. In short, it is an odd sort of leftism that rejects outright the lifetime of financial security a UBI could provide.

Lyons's other claim, that UBI 'makes no sense using a power analysis', is also a fair point, but again, it is generic. All big reforms are hard and, as I discussed in the Power chapter, involve exactly that, the exercise of power. Sure, the implementation of a UBI is going to be hard anywhere in the world. In Australia, for instance, the Greens are the only major party openly advocating it, while both the Labor and Coalition parties are actively hostile. Various social services organisations such as St Vincent de Paul and Anglicare are sceptical, and I can barely begin to imagine the scare campaign likely to be run by the conservative media, who would no doubt home in on the 'dole bludger' angle. As Tim Hollo, executive director of Green Institute, said at the

Melbourne University workshop on UBI, 'big progressive change rarely, if ever, happens by winning arguments, or "convincing" opponents to switch sides. Big change happens when we successfully shift the cultural context and, by doing so, turn an idea from inconceivable to inevitable.' Hollo mentioned free tertiary education, Medicare, the end of the White Australia policy and the apology to the Stolen Generations as examples of big change happening because the political culture, the discourse, had shifted in their favour. Certainly, there is nothing inevitable about the implementation of a UBI, but to abandon it because implementing big change is hard is to abandon politics itself.

So let's now look at what are likely to be the biggest problems UBI advocates face: paying for and implementing a UBI. For me, this is where concerns and criticisms about UBI are best directed and where UBI genuinely has a case to answer. It is also where concerns about political power really bite: marshalling the necessary support for changes to capital distribution to pay for a UBI are going to be hellishly difficult to achieve. Those with a vested interest in keeping tax rates low and continually heading down – which includes a substantial portion of most countries' major business leaders – have shown across history that they are willing to fight long and dirty to relieve themselves of the obligation of supporting government services through taxation. We know from figures released by the Australian Tax Office that fully a third of businesses in Australia pay no tax, and that, on average, the effective rate the rest of them pay is well below the official rate. But this doesn't stop various business lobbies from demanding ever-lower rates. Media investigations such as the Panama Papers and the Paradise Papers have shown that wealthy individuals and large corporations across the globe go to extraordinary lengths to use offshore tax havens and exempt their wealth from taxation. This means that anyone seeking to reform tax regimes to collect the revenue to fund a UBI will face the most ruthless and self-serving opponents imaginable. Such people are unlikely to be argued into the necessary change.

On the other hand, there is evidence that ordinary people around the world are becoming tired of these sorts of arguments and that they are now willing to see higher rates of tax in order to fund better social services. An April 2018 survey in Australia by the Research Now polling organisation found that 64 per cent of respondents wanted 'more public spending on public services and infrastructure, funded by more tax revenue, in particular from wealthy people and profitable companies, and less inequality in Australian society'. A 2017 survey in the United Kingdom by British Social Attitudes showed that 61 per cent of respondents backed higher tax rates in order to better fund the National Health Service. Even in the United States, where opposition to government redistribution of wealth is traditionally high, attitudes are changing. According to Matthew Yglesias, writing in *Vox* ('New poll shows what Americans really think about taxes: the rich should pay more'), polling by the Pew Center in 2017 suggests that 'Americans' top concern about the tax code is that they want corporations and wealthy individuals to pay more taxes. Even among rank-and-file Republicans, soaking the rich is at least moderately popular.' None of this proves outright support for a UBI, but it does suggest that the ability to win people over to such reform is perhaps easier than critics suggest.

The political change necessary to make UBI possible is nonetheless likely to be a project of many years' work and will involve not just winning the arguments about UBI, but shifting the ground in other areas of policy, political practice and, indeed, social attitudes. I doubt UBI is a policy that can just be introduced: I think it will only come about by creating the circumstances in which a significant section of society comes to see it as a viable way of addressing their concerns about security, fairness, equality and personal freedom. As such, UBI should only be proposed in the context of other changes that shift the balance of power in civil society and within our institutions. Too many advocates simply point out the benefits that would accrue from UBI. What they need to do is explain how implementation is possible in the first place. This doesn't mean providing a perfectly detailed and faultless model of a UBI in action – no public policy can ever be specified in that way – but it does mean explaining why we need it and how we might get it.

A key reason for the increasing interest in UBI is that people are concerned that many jobs will be lost to new forms of automation and artificial intelligence, everything from driverless vehicles to 3D printers. In a world of less work, it is argued, we are going to need some other way of paying people a living wage. I share these concerns, but I don't think UBI advocates should use them as the *primary* justification for supporting it. Getting too bogged down in the argument about whether or not robots will take jobs can be misleading (as I argue in more detail in the next chapter). Instead, we should embrace the possibility of less work that arises from the new technologies and rethink our understanding of work more generally. There are compelling social and psychological reasons for making paid work less central to our lives, and the rise of the machines may be the key to embracing such a future.

The more immediate concern with the rise of new technologies is likely to be how they will concentrate wealth, either by machines substituting for human labour or even the *threat* of new technologies forcing wages down, something for which there is already ample evidence (see, for instance, the 2016 report *The Race between Machine and Man* by Daron Acemoglu and Pascual Restrepo). Or it can happen when the technologies create new industries that simply require fewer human workers. Famously, Instagram employed only 12 people when Facebook acquired it in 2012 for a billion dollars, but many high-tech companies employ few people relative to the overall size of their business. Google, for instance, has about 50 000 employees worldwide; Facebook has about 22 000. Writing for the London School of Economics, Clarys Roberts notes that the report on automation that she co-wrote for the Institute for Public Policy Research found that 'jobs with the technical potential to be automated are associated with £290bn of wages per annum. This represents 33% of all wages and earnings in the 2017 UK economy.' In other words, 33 per cent of all wages are vulnerable to being transferred to profits. Roberts offers various reasons why the full amount is not likely to change hands in this way, including the creation of new jobs, but her point is that even if new jobs *are* created, there is likely to be distortions in how they are distributed, with those who lose the old jobs not necessarily getting the new jobs. We already know that there is a trend of wealth away from wages towards profits, and Roberts notes that the 'OECD has shown that over the last four decades, at least half of the decline in the labour share in advanced economies has been due to technological change'. So, concern about technological unemployment by itself is a compelling reason to consider implementing a UBI, but I don't think it is enough. We need to make a more values-based argument as well.

Implicit in the push for UBI, then, is an underlying idea of human happiness and thriving that presumes some very specific things about what it means to be a human living on planet earth at this moment in time. UBI reflects certain values, including notions of equality and fairness, and that is what advocates should stress.

So I suggest the case for UBI is best made by arguing that everyone is entitled to a 'fair share' of the wealth we create through our various human endeavours. All wealth in a society is socially generated, arising not from individual effort alone but from individual effort in the context of knowledge and infrastructure that owe their existence to the wider society. No-one is successful without access to our common physical infrastructures of education, healthcare, and our legal and judicial system, and all these institutions have developed over generations and been paid for by citizens in general. Everyone also shares in the commons, the physical world of natural assets without which none of us could thrive or prosper. Looked at this way, UBI is an inheritance rather than a handout, recognition of our common membership of a polity. As US senator Elizabeth Warren once put it:

> You built a factory out there? Good for you. But I want to be clear. You moved your goods to market on the roads the rest of us paid for. You hired workers the rest of us paid to educate. You were safe in your

> factory because of police forces and fire forces that the rest of us paid for. You didn't have to worry that marauding bands would come and seize everything at your factory – and hire someone to protect against this – because of the work the rest of us did. Now look, you built a factory and it turned into something terrific, or a great idea. God bless – keep a big hunk of it. But part of the underlying social contract is, you take a hunk of that and pay forward for the next kid who comes along.

Once we acknowledge the existence of this 'common wealth', we lay the foundation for a system of UBI that is less a payment due to a hardship – such as unemployment caused by robots or the vagaries of the capitalist economic cycle – and more a fair share of the wealth we have all helped generate as members of society. Similarly, as social activist Eva Cox argued at the workshop, we tend to ignore the contribution the informal economy – the household and the community – makes to our national wealth, and this means that such work is devalued or not even recognised as work. Despite this neglect, Cox pointed out, this informal economy makes the formal economy of paid work possible, and UBI is a way to recognise and reward the contribution it makes.

In advancing this 'common wealth' argument, I specifically include the ownership of personal data as well. Given the primacy of data to the functioning of many of our biggest tech companies, and its absolute centrality to future wealth creation, it is vital that we who generate the data – via our use of everything from online platforms to store loyalty cards, to our smartphones – are compensated for it. If, as many suggest, data is the key commodity essential to the operation of the economy and the ongoing creation of wealth, then it should be subject to a similar royalty payment applied to other such commodities, so that our common ownership of it is fully recognised. UBI is a practical way of recognising that data belongs to everyone.

Whatever the pros and cons of UBI, there is no doubt that citizens throughout the world are rejecting business-as-usual politics. The push for UBI needs to be understood in this context, as a policy response to the uncertainties in developed countries arising from the casualisation of work, the diminishing power of unions to equalise the share of wealth going to profits rather than wages, the consequent increasing inequality in developing nations and the growing intrusion of government – via 'mutual obligations' – into the lives of citizens, particularly those dependent in one way or another on welfare. As a political project, UBI offers a big-picture approach that appeals to people tired of technocratic solutions that have been shown to fail, and a democratic politics – increasingly seen as in thrall to the big end of town – that is unable to confront challenges that are both local and global. But proponents should be careful not to get too bogged down in the minutiae of costings and implementation (as important as they will ultimately be) and thus reduce it to yet another technocratic debate. It should be enough to show that UBI is doable: as economist John Quiggin puts it, UBI is 'challenging but possible'. The concept's growing popularity should be embraced and used as a mechanism for demanding change. Grassroots support is likely to dissipate if the debate descends into a technical argument. This is what happened during Australia's referendum on becoming a republic: the entire debate became one about which model of republic was best and so the 'big picture' goal of independence was lost for a generation.

Let's deal with one final trick critics of UBI use to suggest that it isn't viable, which is to simply nominate an amount for the payment – say, $20 000 – and then multiply that by the number of people in the country and claim that is how much the scheme would cost. This is grossly misleading, as Van Parijs eplains:

> In the context of countries with developed welfare states and tax systems, this does not make sense, because much of the basic income would be 'self-financed' in two ways. [First], it would replace all lower social (assistance or insurance) benefits and the lower part of all higher social benefits. [Second] it would also replace the tax exemptions on the lower income brackets of all households and possibly a number of other tax expenditures – for example, on childcare services or private pensions.

So what does a UBI cost? John Quiggin calculates the cost of a fully implemented UBI at between 5 and 10 per cent of GDP, rejecting outright the claims by Labor MP Andrew Leigh that the cost would reach 23 per cent of GDP. Quiggin nonetheless argues the introduction should be gradual, noting that 'UBI is not a short-term policy option but a vision to be realised over coming decades'. According to Quiggin, a UBI would need to be large before it could be useful. Although a UBI of $6000 a year would cost as much as the existing welfare budget, that relatively small level of payment would not enable anyone to live independently. Quiggin argues that UBI implementation is likely to only be politically possible over several election cycles, and I think that is right. So, rather than beginning with a payment that is universal and unconditional, he argues that Australia could start with 'an income-contingent guaranteed minimum income implemented with a combination of a clawback rate, and a marginal tax rate equal to 40 per cent over the relevant range, and a 40 per cent marginal tax rate on incomes above that level'. Once such a system was shown to have worked, it could be extended to more people.

The risk this approach runs, though, is descending into incrementalism, which falters on the fact that, although it is easier to institute a small change than a big one, it is also much easier to undo small changes. Not only do you slow progress towards the end goal, you leave the gains made in the various stages easy to repeal. A large part of the attraction of UBI is the very fact that it is big and disruptive and upends current arrangements. It actually does change things, and this promise is what gives UBI momentum in the first place. While a gradual approach to implementation may be politically necessary, the end goal of completely implementing a universal and unconditional payment should be kept front and centre.

Political economists Troy Henderson and Ben Spies-Butcher come at the problem from another angle. Their approach involves a stepped implementation of UBI with two initial stages. The first is a redesign of Australia's age-pension system to make it genuinely universal (by removing means testing). The second introduces a new youth basic income (YBI). Once the redesigned pension and YBI are in place, the path to a fully functioning UBI is achieved by gradually increasing the number of people entitled to the pension and the YBI. Essentially, over time, the two payments are extended to more people until, if you like, they meet in the middle, leaving us with a fully functioning UBI. The logic of this methodology is not simply about finding an economic model (which Henderson and Spies-Butcher spell out in some detail) but also about finding a way to build support from as large a political base as possible. Spies-Butcher says that with the model he and Henderson outline: 'It becomes economically feasible for governments to make incremental extensions, say lowering the pension age to sixty or increasing the YBI to thirty, and it creates real constituencies of voters who can see the benefits of the scheme and mobilise behind expansion.'

An alternative funding model is offered by former Greek finance minister Yanis Varoufakis. He proposes what he calls a universal basic dividend. Under this scheme, companies floated on the stock exchange would be obliged to earmark a certain number of shares as commonly owned (in effect, transferring those shares to government ownership), and the government would then distribute the dividends generated by those shares to citizens in the form of an unconditional, universal payment. Some jurisdictions already run schemes like this, the obvious one being the Alaska Permanent Fund, which distributes state oil revenues to citizens. Varoufakis's dividend scheme would extend this idea to all industries. This is my preferred model for funding a UBI because, despite the political difficulties in implementation, it requires less legislative change than a funding model that relies on adjusting tax rates. Despite this preference, I am open to persuasion as to which model is best, and once the underlying principle of UBI is publicly understood, there needs to be a detailed, nationwide discussion of them. Indeed, the topic lends itself perfectly to the sort of citizens' jury, deliberative poll or People's House approach discussed in the previous chapter, and such a public discussion, at that level of seriousness, should be instigated as soon as possible – though it is encouraging to see how many informal public discussions are springing up already, run by everyone from the union movement to local governments, to political parties, to universities and schools.

Let's step back a little and look at the big picture, since UBI is a big-picture approach that is actively pursuing wholesale social transformation. The final section of the Beveridge report says:

> Freedom from want cannot be forced on a democracy or given to a democracy. It must be won by them. Winning it needs courage and faith and a sense of national unity: courage to face facts and difficulties and overcome them; faith in our future and in the ideals of fair-play and freedom for which century after century our forefathers were prepared to die; a sense of national unity overriding the interests of any class or section.

Beveridge was fully aware that such national unity was easier to achieve during a war, but it would be bizarre for us, for anyone, to wish for a war in order to get things done. Not for the first or last time in this book, I will pose the question: do we really need a war for us to take inequality seriously?

As noted above, I have sat in pubs, in church halls, at libraries and at dinner tables discussing UBI with ordinary people, people who aren't part of the political class, who are just voters, and I have heard the hope in their voices as they ask questions about whether UBI could actually work, about how we might implement it, and what the world might look like if everybody really did have a guaranteed basic income. Such grassroots interest has convinced me that UBI is as potentially transformative as its advocates suggest. Ordinary voters recognise UBI for the demand that it is, a demand to smash through business-as-usual approaches and get something done that actually helps them. They recognise that this is not a time for patches but for wholesale, root-and-branch reform, and they smell in UBI the possibility of that sort of change.

And they want it.

One question I was asked at a recent event was this: If you had to narrow it down to one single reason to support a UBI, what would it be?

My answer was: security. I find it very easy to support a scheme that gives ordinary people the sort of security that comes with a lifelong guaranteed income. As such, UBI is more than a more ambitious form of the traditional safety net: it plugs the holes in that safety net and turns it into a solid floor through which we cannot fall. Giving people a basic income is the best way I have heard to ensure that people not only survive as their economic circumstances change – moving in and out of employment, for example – but that empowers them to participate in society more on their own terms. By making the payment unconditional and universal, we remove the indignities routinely dumped upon those forced to avail themselves of the current welfare system. We remove the stigma associated with such social support. We give people the power to say no to bad jobs with shitty conditions and poor pay. We go way beyond the usual calls for a better work–life balance and begin to rethink entirely what we even mean by *work*, a matter I take up in the next chapter.

So here's the thing: having a guaranteed minimum income gives the poor what the wealthy take for granted, the ability to make decisions about the work they do, in and outside the labour market. It says that when it comes to wealth and income, no-one has to start from zero. And so it doesn't just give people more money; it gives them more freedom to choose. Instead of being totally reliant on an insecure jobs market and a wages system that increasingly fails to pay people a sustainable income, that doesn't remunerate the work they do *outside* the formal workplace, and that doesn't recognise the data we currently give away for free, a UBI says we are all entitled to a certain standard of living and a certain level of free choice, *below which we cannot fall*. The universality of UBI reinvigorates the idea that we are all in this together, that there is such a thing as society, that you cannot buy your way out of your social obligations, and that all of us should contribute to, and benefit from, the common wealth.

WORK

> The idea that the poor should have leisure has always been shocking to the rich.
>
> *Bertrand Russell*

If you talk about the future of work, you almost immediately force yourself into the prediction business. It's a bit of a trap, so I'll say at the outset of this chapter that I smashed my crystal ball a while back, my tarot cards are in a box in the garage, and I was never much one for reading animal entrails. Rather than trying to divine the specifics of how technology will affect employment, I will instead consider the way in which work and technology interact and discuss some of the broader implications. We need to get our heads around the idea that technology always changes the jobs we do, and that it always puts pressure on the conditions of our employment. This means that we constantly need to be adjusting our ideas of what 'work' even means, making sure we don't get stuck with outdated ideas that not only trap us in bad practices, but blind us to new possibilities.

One way to understand the world is to look at the words we use to describe it and how they change over time, and the idea of work has a rich linguistic history.

Anthropologists have long noted that some ancient cultures had no word for work. The things they did to survive, like hunt and gather, make weapons and tools, were fully integrated into their everyday lives and often inseparable from play. The citizens of Ancient Greece distinguished between work and labour, where labour was the boring, repetitive stuff that kept you alive and was only fit for slaves to carry out, and work, which was the activity of citizens, the creation of art, philosophy and politics that endured beyond a single lifespan. While the slaves were *labouring*, the citizens of Ancient Greece were busy *working*, inventing Western civilisation.

Jump ahead a few thousand years and an English thesaurus will throw up at least 50 synonyms for work, including chore, slog and grind. Among the antonyms are lazy and failure. Our attitude to work has obviously changed. Our modern work ethic, which we tend to think of as universal and eternal, is nothing more than an artefact of changing social expectations, which are themselves shaped by changing technologies. Today, we pride ourselves on our long hours of work in a way that the Ancient Greeks would find abhorrent and ridiculous. We got rid of slaves, but we internalised the slave mindset, and then industrialisation turned most of us into wage slaves. 'Wage slaves' was a phrase the founding fathers of America used to describe anyone unfortunate enough to have to work for someone else. The citizens of the new republic were meant to be yeomen farmers, as independent of employers as they were of England.

Our modern conception of a work ethic can be traced to the likes of Benjamin Franklin, the American founding father who coined the phrase 'time is money'. For Franklin, the virtue of hard work arises precisely because it is about profit for its own sake as opposed to profit for the sake of purchasing more goods. It is not about self-indulgence and consumption, but about reward for hard work conducted in a trustworthy and thrifty manner. Thus, for Franklin, the pursuit of profit conducted in the right way is the basis of a moral society. An honest, hardworking society is a moral society.

It is fairly easy to see how we go from 'time is money' to 'work is good', to the high priests of finance capital insisting that 'greed is good'. Gordon Gekko's famous line from the movie *Wall Street* captures this transition. He says: 'Greed, in all of its forms – greed for life, for money, for love, knowledge – has marked the upward surge of mankind.' With one simple linguistic trick – substituting greed for work – Gekko reverses the polarity of Franklin's work ethic. Franklin's moral community of a hardworking society becomes Gekko's fractured dystopia of greed and individual elevation. Or Margaret Thatcher's 'there's no such thing as society'. We end up with an ethic that insists more work is always better – because greed is good – until we reach the present day and things have changed so much that the Japanese have the word *karoshi*, and the Koreans have *kwarosa*, and they mean to die from overwork. So we have gone from cultures having no words at all for work to some contemporary cultures discovering that they need a word to describe people dying from doing too much of it. The lives of the ancients may well have been nasty, brutish and short, but finding out we need a word for death from overwork doesn't exactly strike me as progress.

Already, the online world is blurring the idea of what we even mean by work, not just changing the way that we organise it. Platforms such as Uber and Airbnb extract value from once-dormant assets like our parked car or our spare room, potentially turning many of us into taxi drivers or landlords. In the process they are redefining work-related terms like 'employee' and 'worker' and inventing new categories like 'dependent contractor' or 'full-time casual'. I spent a half-day at Uber's Brisbane headquarters in 2017 and watched as office staff read from scripts while they talked to 'driver partners' on the phone, so that they never told those drivers to *do* anything: they only ever suggested or recommended, because to *tell* them to do something would be to imply an employee–employer relationship, and we can't have that because the whole business model would collapse. I asked Sam Bool, the local manager who was showing me around, if it was true that drivers are fired if their customer rating falls below a certain level (generally reported as 4.8 stars). He told me that the star rating is just one tool Uber uses to keep a check on driver performance. Another is the 'telematics' function associated with the GPS tools within the drivers' smartphones. By using those readings, Uber can tell whether a driver is holding the phone in their hand or has it mounted on a stand on the dashboard (as Uber recommends). It can also tell whether the driver has accelerated too quickly, swerved or braked suddenly. Uber justifies this close monitoring of drivers as a way of improving overall safety, which may well be true. But it and the star-rating system are also a way of supervising 'driver partners' without having a formal employee–employer relationship.

The collection of data, too, is breaking down the boundaries between work and play, between us as consumers and us as workers. Data, the commodity on which the whole world increasingly runs, is a key reason artificial intelligence is getting less artificial and more intelligent, and – as I set out in the Media chapter – it is the lifeblood of some of the biggest companies in the world, including Facebook, Google and Amazon. Even companies that we don't think of as tech companies are increasingly becoming *data* companies, and an executive from Woolworths told me recently that Woolies, too, is a data company. Their business is a huge, real-time experiment with prices and stock placement, all of it driven by data that comes from their checkouts, their apps and membership cards, and that is analysed by machine-learning algorithms.

So think about this data revolution for a minute and what it means for our ideas about work. Nearly all of it is generated by we-the-people in the normal course of our daily lives. Every time we turn on our smartphones, do a Google search, share a family photo on Facebook, connect with a colleague on LinkedIn, hire an Uber, have some food delivered by Deliveroo or Menulog, or play a game of chess or League of Legends online, we are, in effect, working for these companies, extracting data from ourselves and giving it to them for free. It's work, but it is not a job.

We tend to treat data as capital, but it might make more sense to think of it as labour. Jaron Lanier, one of the inventors of virtual reality and who works for Microsoft and is an all-round tech dude, points out in his book *Who Owns the Future*: 'Instead of enlarging our overall economy by creating more value … the rise of digital networking is enriching a relative few while moving the value created by the many off the books.' The labour of data creation simply isn't accounted for. The new technologies have not so much destroyed jobs – which is what we usually worry about when we talk about technology and the future of work – as invented a whole new class of work that they have managed to disguise as something else.

This suggests that what work looks like, how we talk about it, and how it changes under the influence of technology, is a lot more complicated than most media discussions admit. The phrase 'the future of work' has become a staple of public debate, and it has come to dominate media discussions on the likely effect on jobs of everything from AI to 3D printing to the so-called internet of things, with #futureofwork becoming a common hashtag on major social media platforms and a familiar tag in bookmarking apps like Pocket. Like many well-used phrases, it distills a series of ideas about a topic of genuine importance, but it has become so reductive, it hides as much as it reveals and risks tipping over into meaninglessness. It has become so misleading that if I could, I would ban the phrase altogether. As I lack that particular superpower, I'll instead explain why we need to find better ways to understand the future of work than are usually embodied in the phrase 'the future of work'.

The main problem is the phrase has come to focus on a single interesting but misleading question: will a robot take my job? The reason for this focus goes back to a 2013 report written by the Oxford Martin School that declared 47 per cent of the 702 jobs examined were likely to be replaced by automation within the next 20 years. The report even ranked those 702 jobs in order of their likelihood to be automated. In one fell swoop, it reduced a complex and detailed area of politics, economics and sociology to a single question – will a robot take my job? – and provided a simple tool that purportedly gave you a way of answering it. The media ran with it, generating all sorts of articles that either simply repeated the *OMG* numbers or looked at where particular jobs were on the list. From then on, every technological advance could be, and has been, slotted into this pre-prepared discussion about 'the future of work'.

The reaction against the Oxford Martin School's take on the robot question was almost immediate. A professor of economics from the Massachusetts Institute of Technology (MIT), David Autor, wrote a paper called 'Why are there still so many jobs?' that set the tone. Everything from Autor's semi-sarcastic title to his insistence that concerns about technological unemployment are nothing new, and his dismissal of such concerns as 'anxiety', not to mention his focus on the limitations of the technology itself, have been aped and emulated in a range of papers, mainly by fellow economists. Autor's paper (and some of the ones it has inspired) makes a valuable contribution to the debate. His empirical work on labour markets across a number of papers is essential reading. But he and others are at such pains to stress that reports like Oxford Martin's are wrong that they end up playing down the changes that are occurring, and they license politicians to ignore the problems altogether. Autor walks the line between dismissal and legitimate concern better than most, but for all the nuance in his paper, he, too, has basically reduced the complex issue of the effects of technology on work to the same question – will a robot take my job? – though his answer is no it won't; or rather, a robot may take some jobs, but new ones will be created. His dismissive tone for those he criticises is displayed in a quote he uses from Nobel laureate Herbert Simons: 'The bogeyman of automation consumes worrying capacity that should be saved for real problems.' While I can understand the professional desire to counter what he considers exaggerations and errors in the general discussion, Autor's contemptuous tone suggests a certainty about predicting the future that is unwarranted. Nowhere is this more apparent than in his confident assertion that computerisation and automation are only likely to be able to replace 'routine' tasks, those that can be precisely written down in a rule-based form. Already, advances in AI, deep learning and neural networks are rendering this conclusion unsafe. Carys Roberts and the other authors of the report 'Managing Automation' note:

> Whereas the technologies that drove automation in the past required clear instructions in controlled environments to substitute for human endeavour, new technologies are now increasingly able to act and problem-solve independently, inferring the appropriate solution or actions on the basis of the external inputs, and 'learning' as they do so. As a result, machines (whether hardware or software) are increasingly able to perform both routine and non-routine tasks, physical and cognitive work. Tasks once thought to be the sole preserve of humans can now often be performed better, and increasingly more cheaply, by machines.

Autor and economists like him are predicting the future as much as the 'futurists' they deride, and as this example illustrates, their presumptions are just as likely to be overtaken by new developments as anybody else's. A little less contempt and sarcasm may therefore be in order.

Other researchers have tried to offer alternative ways of understanding what is happening. So while nearly all the reports looking at automatability, including the Oxford

Martin report, consider the *tasks* associated with given jobs, individual reports vary in how they define and interpret those tasks. The Oxford Martin report, for instance, notes that 'we do not capture any within-occupation variation' and so when the OECD looked at the data, it took a different approach: where Oxford Martin used occupational level data, the OECD used individual level data. This led to quite dramatic differences in the overall results. The OECD report, *The Risk of Automation for Jobs in OECD Countries*, concludes that 'on average across the 21 OECD countries, 9% of jobs are automatable'. In so doing, it contrasts this figure with the Oxford Martin report figure of 47 per cent, saying that the 'threat from technological advances thus seems much less pronounced compared to the occupation-based approach'.

The apparent nuance reports like the OECD's inject into the argument is somewhat illusory. Now, rather than simply argue about whether or not a robot will take your job, the more 'nuanced' question is whether the number of jobs lost will be in the order of 50 per cent (as per Oxford Martin) or 'only' around 10 per cent (OECD). You end up with articles using headlines such as 'Automation threatens 800 million jobs, but technology could still save us' (*The Verge* reporting on a 2017 report by McKinsey & Company), which includes almost farcical comments like this: 'The figure of 800 million jobs lost worldwide, for example, is only the most extreme of possible scenarios, and the report also suggests a middle estimate of 400 million jobs.' So obsessed has everyone become with offering a 'balanced' view of the robot question that the bigger point is lost. Obviously, the difference between 50 per cent and 10 per cent of jobs lost is huge, but surely, either way, we are talking about massive reorganisation in labour markets. Isn't that where our focus should be, rather than arguing the toss on numbers? The net effect of Autor's breezy dismissal of concerns about job losses as technological anxiety, or the OECD offering up the 9 per cent figure as being so much better than the 47 per cent suggested by Oxford Martin, is that it underplays that massive change. 'The future of work' as a concept, as a media story, and even as a subject for highly trained academics and consultants, has become just another sensationalised discussion, and so we really need to move beyond the 'robot question'.

Instead, we should start thinking about work in a fundamentally different way. Automation (from robots to machine learning) is going to change the work we do. Nobody denies that, nobody at all. Jobs will be destroyed, and the argument is really about whether new jobs will be created to replace them. But simply saying 'new jobs will be created' doesn't solve anything either. It just raises a number of other questions about those new jobs: Will there be enough of them? Will they be good jobs, ones that pay well and provide decent conditions of employment? Will all areas benefit equally? What about the changing nature of the skill sets involved in doing them? Will the people who lost the old jobs be able to do the new ones? Will machines eventually replace even the new jobs? Do we really need jobs anyway?

Instead of concentrating on job losses per se, we need to better understand that the revolutionary nature of the new technology lies in how it fundamentally changes the way businesses operate and the effect it will have on how we organise society more generally. It isn't just blue-collar or repetitive, mechanical jobs that are at risk, but white-collar, cognitive work too. I noted in the Media chapter that journalism was the first white-collar profession upended by these technologies, but the ramifications for the professions more generally should be fairly obvious. We are already seeing evidence of changes in areas such as accounting, banking and law. An article in the *Australian Financial Review* ('Professional services is now two very different markets') notes that easier access to information, the rise of independent platforms, and the improvements in artificial intelligence, are all contributing to a fundamental reordering of the legal profession. Paul Rubenstein, managing partner of the Sydney office of law firm Arnold Bloch Leibler, told the *AFR*, 'It's increasingly a buyer's market, where competitive panel arrangements, equal access to information and growing sophistication among clients is driving down the price of professional services.'

Similarly, Australia's major banks recently announced they would lay off 20 000 staff throughout 2018 as technology changes their industry. *AFR* financial reporter James Frost writes that the changing nature of the industry means that the 'solution is automation with digitised functions producing new opportunities for savings and headcount reduction'. The head of banking at KPMG told the *AFR* that 'there was a degree of inevitability about the job losses as more processes became automated'. Some new jobs would also be created in the sector – though not nearly as many as were lost – largely in the area of technology services. NAB (one of Australia's 'big four' banks) announced it was laying off 6000 staff and hiring 2000 tech professionals. The chief executive of Deutsche Bank has said that robots will replace a 'big number' of staff, adding, 'The sad truth for the banking industry is, we won't need as many people as today.' Bank executives I spoke to in New Zealand in March 2018 echoed this: the pressure was on them to reduce staff numbers and use technology instead. In this sense, the banking experience is different from what happened with media, though the same cost pressures brought about by digitisation and technology drive it.

Increasingly, professional firms are offering a bigger range of services, options that would have previously been outside their area of expertise. As the *AFR* notes:

> Law firms like Minter Ellison are moving into non-law consulting work, strategy firms like BCG, Bain and McKinsey are now helping clients carry out their grand plans and the big four – Deloitte, EY, KPMG and PwC – are in everything now, from law through to strategy and consulting work and even advertising.

The disaggregation that is happening in these industries is more a division between 'the high-end work that brings prestige and the big dollars, and … grunt work that clients see as a commodity to be automated or be doled out to the lowest cost operators or performed in-house'. Fiona Czerniawska of the UK-based research firm Source Global Research says that there's 'still demand for highly experienced consultants and lawyers who are capable of innovative thinking and expert problem solving, but that part of the market is quite different to the more run-of-the mill, semi-industrialised work typically done by more junior people'.

Such divisions of labour are indicative of the way in which technology – access to information and digitisation – is affecting the labour market more generally. Mere technical training, whether in technology itself or the specialist knowledge associated with a profession, is increasingly seen as only a small part of the skill set needed to keep people employed. As the World Economic Forum has noted, '[B]y 2020 … social skills – such as persuasion, emotional intelligence and teaching others – will be in higher demand across industries than narrow technical skills, such as programming or equipment operation and control. In essence, technical skills will need to be supplemented with strong social and collaboration skills.' Or as Steve Jobs said in 2011 at the launch of the iPad 2: 'It's technology married with liberal arts, married with the humanities, that yields the results that make our hearts sing.'

Again, the issue is not simply that robots or machines will replace humans but how digitisation, platforms and machine learning change the relationship between humans and the work that they do, and between practitioners and their clients. It is about the way in which unchecked automation or technological change are likely to concentrate wealth in the hands of the few. Partly it is about labour displacement (to use the bloodless phrase of economists), replacing humans with machines. But it is also about the way in which technology changes the demand for high-skilled and low-skilled workers and changes the conditions under which people are employed, and how that then affects the way they live. A 2014 article in the *MIT Technology Review* ('Technology and inequality') uses the example of Silicon Valley itself, the epicentre of the technological revolution, to illustrate what is happening:

> Though California's economy – the world's eighth-largest – is strong in many sectors, the state has the

> highest poverty rate in the country ... About 20 to 25 percent of the population works in the high-tech sector, and the wealth is concentrated among them. This relatively small but prosperous group is driving up the cost of housing, transportation, and other living expenses. At the same time, much of the employment growth in the area is happening in retail, restaurant, and manual jobs, where wages are stagnant or even declining.

What's more, as I noted in the Wealth chapter, the technology companies that generate the most wealth are actually the ones that need the least number of workers. Chris Benner, a regional economist at the University of California, Davis, says 'there has been no net increase in jobs in Silicon Valley since 1998', which is an astonishing statistic when you think about it, and one that flies in the face of all the happy talk about technology creating more jobs. As he notes, 'digital technologies inevitably mean you can generate billions of dollars from a low employment base'. Under such circumstances, you can see why calls by governments and unions and various populists in and around the political class for more jobs, more jobs, is likely to be an inadequate response.

The issue isn't just the number of jobs in aggregate, but whether wage income – the wages we earn from being in paid employment – is an adequate way of distributing wealth. In the period after the Second World War and into the 1980s it was, but the evidence is now very strongly showing that it isn't. OECD figures suggest that 'over the period from 1990 to 2009 the share of labour compensation in national income declined in 26 out of 30 advanced countries' while 'the median (adjusted) labour share of national income across these countries fell from 66.1 per cent to 61.7 per cent'. More and more of the wealth created is going to profit and dividends than to wages, and that is what we are going to have to ultimately address.

Globalisation has been part of the problem too. The cheaper labour that developing countries have been able to provide, combined with efficiencies created by technological advances in communication and transportation, has encouraged major firms to shift production 'offshore', which has greatly enriched workers in developing countries, increasing the overall level of world equality, at least in aggregate. Still, it is worth noting, that claim is somewhat misleading. David Harvey points out in *Rebel Cities* that in India, for example, 'the number of billionaires has leapt from 26 to 69 in the last three years, while the number of slum-dwellers has nearly doubled over the last decade'. The 2018 *World Inequality Report* also takes issue with the idea that globalisation has been a net good. It points out that 'income inequality has increased in nearly all world regions in recent decades, but at different speeds' and, most tellingly, that 'very large transfers of public to private wealth occurred in nearly all countries, whether rich or emerging', which means that while 'national wealth has substantially increased, public wealth is now negative or close to zero in rich countries'.

As Erik Brynjolfsson and others point out in *Foreign Affairs* magazine ('New world order'), the drop in inequality in some developing countries may only be a temporary phenomenon. 'Visit a factory in China's Guangdong Province,' they note, 'and you will see thousands of young people working day in and day out on routine, repetitive tasks, such as connecting two parts of a keyboard.' Such jobs no longer exist in much of the developed world and they are likely to disappear in China too. 'As intelligent machines become cheaper and more capable, they will increasingly replace human labor, especially in relatively structured environments such as factories and especially for the most routine and repetitive tasks.' Foxconn, the Chinese company that builds iPhones, is just one major firm investing heavily in automation, despite the fact that its human workforce is relatively poorly paid (about $400 a month on average for a 12-hour work day). Honorary Chairman Tuan Hsing-Chien said in March 2018 that he expected the total number of employees to decrease from 60 000 to less than 50 000 over the course of the year, with more jobs losses likely.

As the value of technology companies rises, so too does the tendency towards financialisation, or a shift of wealth away from the real economy – where things are made and people are employed – into investments of various sorts, where financial 'instruments' are invented and traded. Economist JP Allen points out in his book *Technology and Inequality* that it is in the very nature of large, wealthy technology companies to concentrate wealth through acquisition and investment. Apple illustrates the perversity of this financialisation. Despite having stockpiled in the vicinity of US$250 billion in cash in offshore accounts (hidden away from the tax regimes of any nation and from investment in the real economy), it has *borrowed* money in order to buy back its own shares. 'Apple … has authorized over $150 billion in share buybacks, spending $10 billion on share repurchases in a recent quarter when their total operating cash flow was $11.6 billion (Apple Insider 2016).'

In a lovely understatement, Allen adds, 'the channeling of wealth to the already wealthy has an impact on inequality.'

Allen notes also that the ability of rich tech firms to stockpile wealth 'confers significant power upon these corporations', and they are therefore able to 'acquire other companies, to hire the most expensive engineers and managers, and to attract the interest of investors in secondary financial markets, such as the stock market'. As he says, 'these acquisitions offer extreme examples of wealth being concentrated into the hands of very few investors and employees', and he gives the examples such as 'the multi-billion dollar acquisitions of very small startups, like Instagram and WhatsApp by Facebook. In each case, hundreds of millions of dollars went to a small team of investors, founders, and early employees.' Entrepreneur Scott Galloway, writing in *Esquire* magazine, illustrates this point by comparison with more traditional firms:

> Procter & Gamble … has a market capitalization of $233 billion and employs ninety-five thousand people, or $2.4 million per employee. Intel, a new-economy firm … enjoys a market cap of $209 billion and employs 102,000 people, or $2.1 million per employee. Meanwhile, Facebook, which was founded fourteen years ago, boasts a $542 billion market cap and employs only twenty-three thousand people, or $23.4 million per employee – ten times that of P&G and Intel.

Galloway also points out that Facebook and Google can generate $10 million in advertising revenue with fewer than 10 employees each, whereas a traditional advertising firm like WPP requires 100 staff to generate the same amount.

In the face of all this, surely we can see that the usual approaches to recouping some of this immense wealth so that it is shared by workers via better wages and conditions is likely to be inadequate. Certainly, one response to all this is to demand better jobs, higher pay, higher minimum wages, and to put in place laws that favour full-time employment over casual, part-time and otherwise precarious work, and I am all for those sorts of demands. This is the logic behind the Australian Council of Trade Unions (ACTU) push to stop employers outsourcing their hiring to labour contractors, who treat workers as self-employed and slot them into what were once full-time positions as casualised individual contractors. As the Secretary of the ACTU Sally McManus has said:

> Employers are able to call people casuals – fake casuals – when they're not casuals. They tell them to go and get an [Australian Business Number] when they're actually permanent workers. They convert them to labour hire just to reduce their wages and conditions. All of these are loopholes that need to be shut down so that we make sure we once again have good steady jobs.

I couldn't agree more, but these traditional responses by unions and others are simply not going to be enough to counter the immense concentration of wealth in the hands of the few. We need comprehensive structural changes if we are going to avoid violence and dystopia.

With all that in mind, I suggest three specific innovations that will allow us to make the most of these technological changes, that will help us think of work in a healthier way, and that will ultimately enhance the quality of life for the many rather than the few: shorter working hours, treating data as labour, and insisting on collective ownership of technology and the businesses that develop and deploy it.

Of these, shorter working hours is not only the most important in the short term, but the one with the best chance of implementation, because it is relatively easy to explain and it resonates with traditional demands workers have made. Unions have a long history of fighting for shorter working hours; in fact, it is currently happening, most obviously in Germany and Sweden. Shorter working hours also dovetails well with other claims such as paid overtime, paid holiday leave, paid sick leave, child allowances, paid parental leave and the like. As Miya Tokumitsu writes in *Jacobin* ('The fight for free time'), these demands are not abstract ones posited by academics for a distant future. They are, instead, 'aimed squarely at reducing profit-motivated working hours and improving workers' self-determination and material conditions. They are tangible, achievable goals that can be built upon.' Cutting the amount of work we do has long been the key way in which we have responded to job losses due to automation, and so the call for shorter working hours is just reinvigorating a well-proven method. The Institute for Public Policy Research lays out the logic:

> If automation raises worker productivity, then it should also raise their wages (in theory at least). With higher wages and/or lower prices, as well as choice over the number of hours worked, individuals can choose to reduce their hours to increase their leisure while maintaining their consumption. If they choose to reduce their hours, the supply of labour would

> fall; partially, fully, or even over-compensating for the fall in the demand for workers in the automating industry. Collectively, society could choose to use the productivity gains of automation to reduce overall working hours.

The claim for shorter working hours is different from the claim for 'more flexible' work hours, though there is some overlap. 'Flexible work hours' is another one of those phrases that hides more than it reveals, in this case hiding that the flexibility is top-down not bottom-up: it's really just a code for business owners to impose more precarious, less secure work, to casualise the labour market, and create so-called zero-hour contracts (where workers are obliged to be on call at the whim of employers) and other ways of turning full-time workers into individual contractors (for example, through the use of labour-hire firms mentioned above). In contrast, the claim for shorter working hours presumes that, by and large, the same pay and conditions will stay in place. The head of Germany's largest union, in launching the push for shorter working hours at the start of 2018, told *IndustryWeek* magazine that 'employees [should] be allowed to switch to a 28-hour week for a two-year period with limited impact on wages'. He said that 'reduced working hours must not go hand-in-hand with a drastic salary cut – for instance when staff are caring for young children or ailing relatives', so part of the union's claim is for 'employers to top up workers' salaries to help make up for the shortfall that comes with clocking up fewer hours'. If this sounds like a pipe dream, it is worth noting that a couple of weeks later, an agreement between the employers federation and the union saw them agree on changes, including a 4.3 per cent wage rise and the ability to reduce their working time to 28 hours per week. This is expected to flow onto other industries.

As with a universal basic income, shorter working hours are not an end in themselves but part of a process aimed at changing the underlying structures of our work, rest and play. To demand fewer working hours is to acknowledge that paid work is only one part of our lives, one part of our identities. So while it is important to make clear the material, social and spiritual benefits that are likely to arise from shorter working hours, the change itself will fundamentally alter the way in which our lives are lived, which means we shouldn't try to be too specific about the ways in which this might happen. The whole point is to free people up and let them choose for themselves what they want to do, and that will include giving them the freedom to do nothing. None of us is truly free until we reach a point where we can goof off and not feel obliged to explain ourselves to anyone for doing it. (The real problem with Donald Trump spending a disproportionate amount of his time as president playing golf compared with previous presidents is not that he isn't working, but that the freedom to pursue pleasure just for the hell of it is not available to everyone else. Our claim should be for us all to be able to skive off in this way.)

The political left can often be their own worst enemy when it comes to work. So much of left-wing politics is connected with the labour movement (rightly and obviously so) that the focus can very easily fall exclusively on work itself and the conditions under which it happens, with no eye to alternative arrangements. It can therefore seem that suggesting people work less, or allowing technology to take over certain sorts of jobs, is antithetical to the goals of the left and the labour movement. The left have also tended to be spooked by right-wing scare campaigns that seek to characterise claims for better pay and conditions as evidence of laziness. This leads them to sometimes overcompensate and insist on paid employment as the defining characteristic of a good life. We saw an example of this with Australian Labor leader Julia Gillard, as prime minister, extolling the virtues of hard work at the expense of nearly every other personal value. Even when she highlighted the value of education, it was largely in service of getting a better job. 'When my parents migrated to this country they didn't come asking for a free ride', she said during an election speech in 2010,

> they came seeking a fair go, and they found it. They found it and they worked hard for it … My mother always worked as well and because of their hard work my sister and I were able to have a future … I've been talking about jobs all of this election campaign because I believe in work. I believe in people having the benefits and dignity of work. I've believed that all of my life.

This is wrongheaded but in such a taken-for-granted way that we barely notice. Dignity is not conferred on people by 'hard work'. Hard work and the way it is done *reflects* human dignity, it doesn't create it. So human dignity is enhanced not just by work, but by any number of other activities, and these are ultimately swamped or ignored when we insist that work is the only path to human dignity. Espousing such values – and Gillard is hardly the only person on the left to do so – makes it difficult to mount a case for shorter working hours because 'not working' is seen as diminishing dignity. It's crazy. As Richard Denniss says in his book *Curing Affluenza*, 'Most of us could work shorter hours, and it would benefit all of us to do so. Our decision not to do so is as much cultural as personal. It has nothing to do with economics.' Our work ethic is still geared around the demands of an industrial economy of scarcity, rather than the service and information economy of plenty in which we increasingly live, and we have to unlearn the habits and expectations of that outdated mindset. The politics of achieving shorter working hours, therefore, is likely to begin with the left and the formal labour movement rediscovering – or re-emphasising – the need for working people to work less. And there is some evidence that they are starting to do this. We have, for instance, seen the rise of left-wing think tanks such as Autonomy in the UK that are dedicated to investigating the sort of postwork agenda I am outlining here. Even more importantly, the British Labour Party is adopting some of these schemes as party policy, arguing for an increase in the number of public holidays (as well as examining different forms of business ownership).

The benefits of shorter working hours are genuine and undeniable. It would mean more time for other things, including friends and family. It would allow us to pursue interests or hobbies, further education, other paid work if we so chose, or even just to sit around and read a book or watch telly, but it would also give us the time as citizens to involve ourselves in community activities, including politics. We could volunteer at our children's schools or our parents' retirement villages or the local football club. Beyond these most obvious benefits, there are also benefits for childcare. The current high demand – and therefore cost – of childcare is almost entirely a by-product of long working hours, and so a reduction in the overall hours we work would take the pressure off these services and lower the costs involved. Still, I stress again, the idea isn't to specify in advance what people might do if less of their life was devoted to formal work, but about creating the conditions so that they can choose for themselves.

Overall, I think the four-day week is preferable to the six-hour day as a starting point, because of the environmental benefits it brings in reduced travel: it is one day less a week in which you have to travel to work, with the commensurate reductions in transport usage. Whatever the initial goal, the point would be to constantly revise and, as possible, reduce hours even further. We need to build into whatever scheme we devise the presumption of ongoing reductions in work hours. We may never get to zero or genuine postwork, but it is important to maintain that as the overall goal. Otherwise, we risk slipping back into habits of the recent past, where reduced working hours was lost as a goal of policy and activism, thus allowing working hours since the 1990s to gradually increase. Shorter working hours needs to become normalised, not a one-off.

As well as working less, we need to recognise the way in which our leisure time has increasingly been coopted into creating wealth for others and how this means we have to think about work differently. For all the arguments about technology creating new jobs, we have tended to ignore perhaps the biggest and potentially most lucrative job-creation scheme to come out of online platforms: the idea of data as work. As I've noted, we treat data as capital, a resource that is gathered by companies and organisations and monetised (often though advertising), in return for which they let us use their products (Google, Facebook and the rest) for free.

As Ibarra and colleagues, the authors of the paper 'Should we treat data as labor?' argue:

> DaC [data as capital] treats data as natural exhaust from consumption to be collected by firms, while DaL [data as labour] treats them as user possessions that should primarily benefit their owners. DaC channels payoffs from data to AI companies and platforms to encourage entrepreneurship and innovation, while DaL channels them to individual users to encourage increased quality and quantity of data. DaC prepares for AI to displace workers either by supporting UBI or reserving spheres of work where AI will fail for humans, while DaL sees ML [machine learning] as just another production technology enhancing labor productivity and creating a new class of 'data jobs'.

Some authors of the paper work for Microsoft and so tend to see the matter through the lens of maximising profit for that company. For instance, they see a potential advantage in paying users for data inputs (using Microsoft Bing searches rather than Google, for instance) as a way of competing with the huge number of free users available to Facebook and Google, where ultimately the data is used to train various artificial intelligence and machine learning programs. So, they point out, companies like Microsoft 'lag Facebook and Google in the data race to train ML systems. Returning more of the gains to data laborers might help them compete in creating AI systems.' And, indeed, companies directly paying users might be one of way of doing it.

But as I argued in the chapter on wealth, a more efficient, and fairer, method would be for governments to extract payment from the corporations via taxes or some form of capital account system and then distribute that to citizens as some form of universal basic income. Aggregating the payments in this way not only reduces the complexity of firms entering into employee–employer relationships with individual users, it also removes data from a market-based arrangement and thus diminishes any tendency for people to get competitive by providing more and more data. It also diminishes the discomfort that may arise from people selling their personal information directly for money. Data as labour recognises the contribution that people are making to the common wealth, it provides some meaningful return on the activity itself and thus raises its status, without turning the whole thing into yet another competitive labour market. Effecting payment for data via a UBI also seems a more efficient way of ensuring an adequate return on labour than the suggestion, made in the paper, that users form data unions. Distributed as a UBI, it would also address disparities that arise from access to technologies in the first place.

As John Quiggin writes in his essay 'Peak paper', the information economy lets us 'abandon the 20th century social model in which adults spend most of their days in an organized workplace' because 'much of the value in the information economy is generated by informal interactions through various forms of social media'. He argues that if we combine 'this trend with steadily increasing productivity', it is 'possible to envisage a massive reduction in formal working hours, perhaps to the 15 hours a week envisaged by Keynes nearly a century ago'. We are all working for these tech companies for free by providing our data to them in a way that allows them to hide our contribution while benefiting immensely from it. It is way past time that we were paid for this hidden labour, potentially using that income to offset reductions in our formal working hours.

Wherever you stand on the 'robot question', whether the new technologies are launching us into a jobs apocalypse or a period where many new jobs are created – or some point between – there seems little doubt, as shown throughout this book, that the trajectory we are actually on is towards more insecure work and greater concentration of wealth. We are fulfilling the prediction made by technologist Norbert Wiener back in the 1950s: 'Let us remember that the automatic machine ... is the precise economic equivalent of slave labor. Any labor which competes with slave labor must accept the economic conditions of the slave.'

Under such circumstances it becomes essential – if we want fair societies and functioning democracies – that those doing the work share in the ownership of companies producing the wealth. The matter of who owns the means of production is, of course, as old as capitalism itself, but it is brought into ever-sharper relief as the new technologies and globalisation change work in the ways that I have been describing. The 2018 *World Inequality Report* provides further devastating evidence of the trend towards international inequality, and it notes one of the key causes of this is 'the unequal ownership of capital', saying that 'since 1980, very large transfers of public to private wealth occurred in nearly all countries, whether rich or emerging'. But this is not simply a redistribution problem. It is one in which we have to reimagine our role as workers and citizens.

So we don't just have to campaign for worker ownership and control of firms, we also need to start thinking about the new technology in the way that owners do. Writing on the Co-op News website, journalist and academic Nathan Schneider notes that ownership is 'the ground where the tug-of-war for the next social contracts is being played. Who owns what will determine who really benefits.' But, he adds, owners also 'decide which tasks to invest in automating and what happens to the people who used to do those tasks'. Instead of being victims of technology, he says, workers-as-owners need to become shapers of it, and ownership of firms allows this to happen. 'Rather than fearing how machines might take work away, workers can imagine how they could use those machines to make their lives easier – in ways better and fairer than the investor-owners would.' Transforming this point of view could extend to the technology itself, and Schneider notes that the AI developed by a worker-owned business 'may be intelligent in ways the investor-owned counterparts can't be'. Worker ownership is therefore another way of changing how we think about work, how we think about technology, and – as with the ideas of the commons, sortition, a universal basic income, shorter working hours and data as labour – not just a means to a fairer society, but a *demand* that fundamentally changes how we understand the creation and distribution of work and wealth.

To build a fairer and more equitable society, we need to make the basic institutions of the state and economy fairer and more equitable too, and that means developing ways to change the ownership of those institutions and the way power operates within them. Again, this isn't just about changing who owns what, but about building into firms the logic of fairness and equity we want to see extended to society as a whole. A report by the Institute for Public

Policy Research's Commission on Economic Justice puts it this way:

> The aim of ownership reform should be two-fold: to give more people a share of capital, both as useable wealth and for its income returns; and to spread economic power and control in the economy, by expanding the decision rights of employees and the public in the management of companies.

It is not as if the concept of worker ownership of firms, of mutuals and cooperatives, is some weird idea that no-one has ever considered before. All of these have a long history, with many examples of them working very well. The trouble is there are a number of structural and cultural constraints placed on expanding and normalising the idea, which need to be addressed. The IPPR report argues that the main barriers to expanding the sector are a lack of access to capital, the existence of legal disincentives, and a 'wider business ecosystem' that is 'poorly equipped to support the co-operative sector'. None of these is insurmountable, and they lend themselves well to the sort of community organising I outlined in the Power chapter. They are exactly the sorts of policy issues a People's House could prioritise. They suggest a way forward for a union movement trying to reinvent itself for the modern economy. Speaking of the future of unions, Per Capita research fellow Tim Lyons writes in *Eureka Magazine* that the 'answer to the challenges of work is to rebuild solidarity of purpose and action. A solidarity unfettered by legal structures of collective bargaining but anchored in human relationships and, yes, in rational self-interest.' This is exactly right, and it seems to me that worker ownership of firms fulfils the brief perfectly. I would love to see unions adding worker ownership to the list of demands that informs the invaluable work they do more generally, and I think it would go a long way to reinvigorating their role, helping make them more attractive to the younger, less organised workers of our increasingly precarious workplaces.

In talking about work and 'the future of work', then, we not only need to get beyond the robot question, we need to get beyond our usual understandings of what work is and how we might earn the income we need to survive. The inequality that we thought had been solved in the period after the Second World War has returned with a vengeance, and it is built into the fabric of some of our most important public and private institutions. So while there is immense room to improve the pay and conditions of average workers in the short term through industrial action, I think that, looking ahead, unions, labour parties and everyone else interested in the general welfare and life chances of ordinary people can't just focus on wages and conditions, as important as they are: the solution will only be found in changes to the ownership of capital and the way wealth is distributed more generally. You don't have to posit a jobs apocalypse on the back of new technologies to insist that something needs to be done. And you can't just blithely say 'but the technologies always create new jobs' as if that solves current and ongoing problems. We are at the point where inequality is threatening to undermine not just conditions at work but the whole logic of representative democratic government and, therefore, a functioning society.

Shorter working hours, fair reward for the data that increasingly drives the economy, and worker ownership and control of firms need to be at the heart of our demands. But in all this, let's not lose sight of the overall goal, which is a more secure and equal world in which we can all realise something approaching our full potential. Central to this sort of transformation is likely to be a different approach to education, particularly higher education, to reimagine it to prepare us better for the work – and postwork – of the future.

EDUCATION

> How alien these present-day categories would have been to the minds of the founders of the republic can perhaps best be seen in their attitude to the question of education, which was of great importance to them, not, however, in order to enable every citizen to rise on the social ladder, but because the welfare of the country and the functioning of its political institutions hinged upon education of all citizens.
>
> *Hannah Arendt,* On Revolution

We ask education to bear a lot of weight in our societies. In many ways it is held out as the fail-safe measure most likely to deliver us from doom, our first and best resort to save the world. This has been the case for most of the industrial era, and it is only likely to intensify as we enter the next phase of the technology revolution. Need a better job? Get a better education. Want a well-functioning democracy? Educate more citizens. Want to make inroads into income equality? Break down the barriers to education for people of all classes. Want to make sure women have access to the same opportunities as men? Make sure women are well represented in STEM (science, technology, engineering and maths) subjects at university. Want to increase wealth in developing nations? Educate.

Education, we generally believe, makes us good citizens, well-rounded human beings and employable workers, and indeed, there is strong evidence for all of this. A person with a higher education degree, on average, earns twice as much as someone who doesn't and is less likely to find themselves unemployed: economist John Quiggin points out on his website that during the deep recession of 2011, the unemployment rate among degree holders held steady at 2.5 per cent. As the nature of work has changed over the last century, we have asked education to be available to more and more people, and that they stay in it longer and longer. In Australia, for instance, we now benchmark educational achievement at school level against completion of year 12, and the numbers doing that have risen from 47 per cent for baby boomers to 63 per cent for gen X and 75 per cent for gen Y. The government's current target is that 90 per cent of us finish year 12, and the figure currently sits at 88 per cent. More and more people are expected to achieve some level of higher education too, and the number of people doing that has increased dramatically. The Australian Bureau of Statistics says that 'Nearly two thirds of Australians aged between 20 and 64 years now hold a non-school qualification. Since 2004 this proportion has increased from 56 per cent to 66 per cent.' This is a worldwide trend and the OECD notes that, among member countries,

> Between 2000 and 2012, the proportion of people without upper secondary or post-secondary nontertiary education has shrunk at an average annual rate of about 3%. Meanwhile, tertiary

> education continued to expand during the same period, growing more than 3% each year. For the first time, in 2012, about one in three adults in OECD countries held a tertiary qualification.

We invest huge amounts of public money in education (and I use the word 'invest' on purpose) and, on average, in OECD countries, governments devote 6.1 per cent of GDP to all forms of education. Above and beyond that is the money private citizens are willing to invest in education, and using Australia as an example again, to put a child born in 2018 through private school from kindergarten to year 12 is going to cost you just shy of half-a-million dollars, and that doesn't include extras like overseas travel, sport or specialist subject tuition.

But let's take a breath.

You don't need this barrage of statistics to convince you that all societies value education and that our devotion to it is only likely to increase. Still, we debate how education should be funded, what it should be trying to achieve, the value of early education programs or the need for various postgraduate degrees. Beyond that is endless debate about pedagogy itself, the way in which we teach, and that can include everything from the pros and cons of phonics to the efficacy of methodologies such as Montessori and Waldorf (Steiner), to the costs and benefits of home schooling and online learning via the rise of MOOCs (massive open online courses). Although it is well beyond the scope of this chapter to address these many specialist debates, we will confront some of the mainstream presumptions about education, particularly its relationship with work. We are entering a phase where the push is on for 'lifelong learning', a response to the changing nature of the labour market. As a principle it has a lot to commend it, but the presumptions behind it also indicate the pathologies of our work-obsessed societies, and they have a huge bearing on what we even mean by education.

In all the discussions about the impact technology will have on the jobs we do – whether that is framed in the reductive terms of 'will a robot take my job?' or in more sensible terms about the structural transformation of workplaces and societies – no-one, absolutely no-one thinks that we are in for anything other than massive change. But here's the weird thing: even as people concede that such change is coming, few actually acknowledge that this means the world will look very different to the way it does now. Most of the discussions of 'the future of work' – and this includes discussions about the future of education – ultimately presume that there will still be workers who want jobs, employers who provide them and institutions that train them. They presume that our main goal in life will still be to get a job or to retrain for a new one. Despite acknowledging the change that is coming, they spend their energy on trying make that change not change much at all, and this presumption is misguided in the extreme. The way in which technologies like 3D printing, artificial intelligence and the various online platforms will affect the economy is not just a matter of taking or creating jobs, but of changing fundamentally the structure of society. I want to spell this out a bit so as to make clear what I mean by structural change and why we cannot simply presume that the institutions – like education – can simply go on as before. So let's look at an up-and-coming technology that illustrates this point: autonomous vehicles.

Back at the dawn of the automobile era, predicting that a string of petrol stations would spring up to allow cars to refuel was a pretty common and safe guess. What far fewer people predicted was Bunnings, or Walmart, or other big box stores that only became viable with the spread of cars and the roads they drove on. So while there is no doubt that jobs will be lost as transport goes driverless – truck driving, for instance, is the single biggest employer of men in the United States – the ramifications of the technology will ripple out much further than that. To understand that, it helps to think of them not as 'driverless cars', which conjures notions of horseless carriages, but as autonomous vehicles, or AVs. Then we can imagine them not just without drivers but also without steering wheels or indicators – why would they need either? – and with any number of other enhancements including desks, screens and even beds. There is no reason on earth why an AV has to look anything like a car of today: they are likely to come in all shapes and sizes, purpose-built to fulfil specific functions, including as a multifunction workspace. A fleet of driverless vehicles is not just a fancy taxi service, but one that fundamentally transforms transport. And when you do that, you fundamentally transform cities and the built environment more generally. So consider what that means.

Electric AVs have far fewer working parts than cars on the road today, 100 times less in fact. This means less wear and tear, and longer lasting vehicles. Less wear and tear means less maintenance and therefore fewer mechanics. We won't need petrol stations anymore, and recharging is likely to be done at home or through outlets in our already existing network of power poles around town, maybe even directly via solar panels embedded in the road itself or by simply swapping out batteries at automatic change stations. AVs are much, much safer than vehicles under the control of humans, so the number of accidents, deaths and injuries will plummet. This will have an effect on health costs, but the biggest effect is going to be felt by insurance companies. The size of the US insurance market is $200 billion per annum and it will likely collapse in the same way ad revenue did for media companies. I have spoken with insurance executives who are already examining new business models.

AVs don't need road markings or traffic lights; indeed, they don't need lights at all, and it isn't hard to imagine – or maybe it is – how much quieter cities will be with electric rather than internal combustion engines. What possibilities does that open up? What will be the knock-on effect on, say, double-glazing? In a world of AVs that you can summon on your phone with relative ease and reliability, it becomes viable to simply not own a car. KPMG already predicts car ownership will drop by about 80 per cent as people opt for drive-sharing services. Cars will go the way of movies and CDs: people will no longer own the physical object but will instead rent them; cars will go from being a product to being a service. In that world, AVs will circulate the city waiting to be hailed by an app. They will not need parking lots, certainly not the multistorey sort we are used to at the moment, and so we will be able to find other uses for all that space. What happens to motels when we can sleep in our car overnight as it drives us to our destination? What happens to offices when our car can be fitted with a desk and wi-fi? And, of course, whither Bunnings and Walmart?

Major technologies, once adopted at scale, always bring about this level of transformation, but it remains difficult for us to imagine ourselves in that altered space. So even as we contemplate change, we hold in our heads the hidden assumption that we are transitioning to a world that looks a lot like the one we are in now, or even, to one that looks like the one many of us think of as the golden age of the 1950s and '60s. When governments promise 'jobs and growth', and when businesses demand that governments provide training for the 'workplace of the future', the future they are imagining looks a lot like the one we think of as 'normal' now. That is one of full employment, of 'good' jobs and an ongoing career, of eight hours a day each of work, rest and sleep. Even unions fall into this trap, and are often more interested in protecting current practices – or ideal practices of the immediate past – than confronting any sort of more fundamental change.

So here we are on the verge of massive change in what work is and how wealth might be created, in the early stages of technological revolution that can potentially produce a society in which the need for ordinary people to spend most of their days – indeed their lives – working for other people and creating wealth in which they do not share fairly is no longer the norm, and nearly everyone thinking about this stuff is simply presuming that the basic shape of society will remain the same, that there will be workers, there will be bosses, and that the point of education is to get you a job. It's as if a world full of AVs is one in which the driver is merely replaced with a computer program rather than one in which everything associated with transport, from shopping to holidays to work, is fundamentally altered.

To reduce education to something you do to prepare for a lifetime of work, as skills training for a job, is to reduce life itself to work. Even where people stress the need for a rounded education, for people skills and other interpersonal capabilities, if the call is only to develop those in order to be 'job ready', then we are not just missing the point of education, we are defining life itself as the jobs we do. As ever, part of the problem is the all-pervading influence of the logic of neoliberalism, and education has become as embedded in that logic as any other institution. Under this sort of thinking, education is just another product. Linguist and political commentator Noam Chomsky notes that

> [e]ducation is discussed in terms of whether it's a worthwhile investment. 'Does it create human capital that can be used for economic growth?' It's a very distorted way to pose the question. Do we want to have a society of free, creative, independent individuals able to appreciate and gain from culture achievements of the past and add to them, or do we want people who can increase GDP. These are not the same thing.

Once you presume education is a product designed to increase GDP, it is only a short jump to thinking that what is needed is more competition in the sector, with all the concomitant ideological and social baggage that goes with that approach.

So universities themselves have become top-heavy in administration to compete with each other in climbing up international rankings, a large part of which is about attracting students from overseas. 'In this market-sensitive environment', writes Jennifer M Gidley in her book *Postformal Education*, 'university administrators stripped back their core business, sacrificing academic tenure, non-commercial research, and non-commercial disciplines, eg, humanities. Higher education institutions, old and new, now compete for market share in the market economy.' In Australia, for example, education has become the country's third-largest export earner, bringing in $21.8 billion in 2016 in fees from international students, mainly from Asia. Kristen Lyons and Richard Hil point out in *The Conversation* ('Vice-chancellors' salaries are just a symptom of what's wrong with universities') that 'universities run multi-million dollar marketing campaigns to "strategically differentiate" themselves in order to gain greater "market share"'. University vice-chancellors (VCs) look more and more like any other corporate CEO, right down to their burgeoning remuneration packages, even as the workforce they preside over becomes more precarious. Of all the professions – with the possible exception of journalism – academic jobs have become the most unstable, with huge increases in the number of part-time (sessional) workers and a diminishing number of those who can pursue teaching as a career. VCs, meanwhile, thrive, and in Australia earn an average of $890 000 per annum. Twelve of them receive more than $1 million a year, with the best paid receiving $1.4 million, which, as Hil and Lyons point out, means they receive more pay in a week than a casual employee receives in a year.

The neoliberal influence in higher education is nowhere better illustrated than in the rise (and fall) of for-profit providers. Around the world, private organisations were encouraged into the sector – often through the use of government-funded incentives – and their failure has been almost universal. Whether it was a private provider like Edison in the United States, or the introduction of vouchers in the Swedish education system, or the virtual privatisation of the vocational education and training sector in Australia, the pattern has been the same: meteoric rise followed by collapse. In Australia, the Productivity Commission says, 'Reforms to the vocational education and training (VET) sector illustrate the potential for damaging effects on service users, government budgets and the reputation of an entire sector if governments introduce policy changes without adequate safeguards.' (Which is what bureaucrats say when they can't use the word clusterfuck.) Economist John Quiggin is more straightforward in his submission to the Senate Education and Employment References Committee inquiry into vocational education and training in South Australia. He writes that vocational education in Australia is in a state of crisis.

> Traditional models of on-the-job training (apprenticeships and traineeships) are in decline. Funding for vocational education through the Technical and Further Education (TAFE) system has been slashed leading to the closure of many TAFEs and large-scale loss of teaching staff. Meanwhile, billions of dollars have been wasted on ideologically driven experiments with market competition and for-profit provision.

When you treat education as a product, it is not surprising that you then see it in almost purely instrumental terms as well. Technology is part of the change, and the shift to online learning continues apace, seen as a way of increasing revenue by decreasing jobs. Students are no longer students but customers; subjects deemed to have little commercial value are discontinued, and courses in English, history and ancient languages disappear or morph into more corporate-friendly subjects such as communications. Courses are increasingly designed around the requirements of industry (even as industry struggles to articulate what exactly those requirements are), and credentialism and credential inflation are rampant: that is, those with higher degrees artificially raise the requirements for all applicants, by seeking jobs that don't necessarily require a degree.

Any wider value in learning – and a more expansive view of what it means to be a citizen, to be something more than an input into an economy – is lost. A classic illustration of this is a 2017 report by the Business Council of Australia (BCA). In many ways it is an admirable document that outlines good and necessary reforms. Its suggestion, for instance, that vocational training and university education be treated as a single higher education sector makes a lot of sense, as does its insistence on breaking down the stigma (relative to university) associated with VET learning. (The report points out that a nurse trained at a university is likely to earn more over a lifetime and be accorded a higher status than one trained at a non-university institution.) There is also value in BCA's argument that government needs to develop new datasets to help prospective students better access the information they need to make good decisions about what and where to study, and that this new data be brought together in a single online portal that allows students to easily navigate the myriad options available. I am also at one with BCA in its demand that we recognise multiple ways of learning, that not everyone is academically inclined. 'It is imperative', the report says, 'that we ensure students who are traditional learners and academically successful are offered a learning environment where they can be challenged and thrive. However, it is equally important we value other forms of intelligence and offer those students environments where they can also be challenged and thrive.' Amen to that. BCA is also correct to recognise that technology is likely to change the work that we do, which is likely to mean the need for retraining and reskilling, and thus its focus on lifelong learning (though I have concerns about this, which I will come to). Under such circumstances, BCA is right to suggest that higher education, in particular, is likely to need to be more flexible, that employers are likely to have to provide more on-the-job training than they do now, and that higher education will need to be provided on a more modular basis – shorter, non-degree courses. After all, if the future of work is likely to mean that we change jobs and even shift careers on a more regular basis, we cannot simply return to higher education and do *another* full degree every time we need to reskill.

Despite this recognition of the changing nature of work, the report nonetheless falls into the change-no-change trap I mentioned above, and this failure of imagination is reflected in the title of the report itself: *Future Proof: Protecting Australians through education and skills.* It conjures not a world of possibility and achievement, let alone one of happiness and fulfilment, but one of fear and uncertainty, where education is seen as a prophylactic, an astronaut suit we slip into so we might survive the hostile environment of 'the future'. For all its welcome words about 'greater equality of opportunity, where anyone – regardless of their upbringing – has every chance to realise their potential' and that 'education should develop free, creative and independent individuals', the report ultimately sees education as jobs training with our role as citizens being to engage in a life of work. The freedom and individuality it talks about doesn't extend much beyond our freedom to choose a job. Sure, it says, your education might have some other value, but it 'seems counterintuitive to suggest that people who add to the cultural achievements of the past cannot also add to economic growth':

> There is no doubt that young people need their education to teach them to think, be independent individuals and ensure they have foundations to become a good citizen, but that is not enough. Young people today also need their education to prepare them for a more skilled labour market and prepare them for the world of work.
>
> ... While some of this mismatch could be managed through workplace training, the reality is that our universal education system ... *must have a purpose of preparing people for work.*
>
> ... Australia should take the wisdom of philosophers, educationalists, intellectuals and public commentators who have pondered the role of education, *but* we should also bring a modern context that *focuses on the world of work* and the fundamental role education now has in preparing people for work. (My emphasis)

In the end, BCA asks, almost in despair, 'What is the purpose of all these additional years of education, if it is not to prepare people for work?'

So notice what is happening here: at the very moment that everyone agrees that the nature of work is fundamentally changing, and the very idea of work as the key to self-worth and self-fulfilment is being challenged by its changing nature – particularly for blue-collar males – such reforms to education are doubling down on an approach that sees it almost exclusively in terms of getting people ready for work. As Cathy Davidson, professor of English and director of the Futures Initiative at the Graduate Center, CUNY, writes for *The Guardian* ('We must reverse the "outcome oriented" educational monster we have unleashed'), no generation of students 'has faced greater global challenges or an educational system more in need of redesign to prepare them for these challenges'. So while students 'are warned that the "robots are coming"' they are nonetheless 'offered an educational system ... designed not to combat the robots but to turn students into poor facsimiles' of past generations.

It has become increasingly common to hear that the issue isn't that robots will take our jobs but that humans will increasingly work alongside robots or will use other forms of technology – such as AI-powered personal assistants – to complement the work that they do. This is no doubt true, but it also has the ring of amelioration about it, something that people say to acclimatise us to technological change as non-threatening. Again, there may be some value in the approach, but it can also lead to ridiculous claims, such as the idea that we should mandate human drivers be available in driverless vehicles (as has been suggested by legislators in Florida). But the less discussed aspect of us working with robots and various forms of AI is that in so doing, the jobs we do will become deprofessionalised and deskilled, a trajectory completely overlooked by those who insist that education is almost entirely about giving us the skills necessary to do the jobs of the future. In the Media chapter I spoke about the way the rise of blogs and the Google search engine allowed an influx of amateurs into the professional field of journalism, and the same thing is happening in nearly every other profession. In supermarkets, a single employee can supervise a wall of automatic checkouts that are being operated by customers. Uber doesn't need to train drivers in the layout of any given city because they all have apps that can tell them where to drive. Increasingly, the work traditionally done by young lawyers – searching case precedent and other research – is being done by software. Small businesses can manage their own tax returns using apps such as Xero. A single architect using software to design a range of different solutions can handle multiple building projects using modular construction. And, as the head of a Melbourne architectural firm told me, increasingly, that manager doesn't need to be an architect. At the very least, the skills needed to thrive in this new environment are changing, and the ones becoming more necessary are the soft skills associated with arts degrees, the very subjects universities are abandoning and governments downplaying.

To the extent that reports like *Future Proof* acknowledge the value of education beyond its utility in producing workers for the economy, a certain level of attention is paid to giving people choice and encouraging creativity. They join those who say we should pursue work that we love, that we should find our bliss, and that if you 'find something you love to do, you'll never have to work a day in your life' (a quote without clear provenance, but one that is widely used). The BCA report – which I am singling out not because it is particularly bad, but because it is entirely typical – makes this notion fundamental to its approach to education. It says that 'that when people are considering tertiary education they start with two questions … What am I good at … What do I like/am passionate about/am inspired by?' These are the first steps in a process that is meant to lead you to a *job*, the next ones being to identify the 'different roles available in different industries that would suit these strengths and interests … The VET or HE courses that are relevant to these roles … The professional bodies that are relevant to the particular industry or role, and which courses they accredit.'

Now, the idea that we should pursue work that pleases us is, on one level, good advice – why, if you could avoid it, would you do anything else? – but it is also loaded with presumptions that are often left unexamined. How, for instance, are you meant to discover what you love in the first place? The idea that 'what we love' simply emerges in the course of our lives is more problematic than it seems, a fact that was brought home to me by my own son. When he was 12 years old, he became interested in classical dance and so started having ballet lessons. After a few lessons, the teachers informed us that he had some talent in the area, and Noah himself was obviously enjoying the classes. After about six months, he asked if he could audition for the Victorian College of the Arts Secondary School (VCASS), a specialist arts school that offered a full-time dance program along with a regular academic education. Noah auditioned, was accepted, and was at the school from years 9 to 12. After graduating, he successfully auditioned for the Australian Ballet School, and three years after graduating from there, took up his first contract as a professional dancer.

The amount of hard work involved in this process, over seven years, is hard to describe, but I don't doubt for a second that he thought it was worth it. During a chat with him about his years of training, I made a passing comment about how lucky he was to have found something he loved, and how that love had made it possible for him to maintain the sometimes gruelling schedules involved. 'No, Dad', he said in no uncertain terms, 'that's not how it works.' Didn't I remember, he asked me, that he had almost quit in year 9, his first year at VCASS? I remembered the discussion very well, and was incredibly proud of how he had assessed the situation. He had started two years behind everyone else and thought he would never get to the same level of proficiency. After much contemplation, he decided to stick at it, and he set himself an even more demanding schedule than that being pursued by the others in his class, getting in early each morning, and making himself available for every competition he could enter. As he explained to me, 'I didn't work hard because I loved it. It was only when I started working really hard that I discovered I loved it.' Telling kids to do what they love is often glib advice that overlooks the process that might be involved in actually working out what that is.

Nonetheless, the mantra persists. So pervasive has this call become that academic and author Miya Tokumitsu has reduced it to an acronym, DWYL – do what you love. In a series of articles for *Jacobin*, and in her book *Do What You Love: And other lies about success and happiness*, Tokumitsu argues that the problem is that the DWYL advice 'leads not to salvation, but to the devaluation of actual work, including the very work it pretends to elevate – and more importantly, the dehumanization of the vast majority of laborers'.

'Superficially', she says, 'DWYL is an uplifting piece of advice, urging us to ponder what it is we most enjoy doing and then turn that activity into a wage-generating enterprise. But why should our pleasure be for profit?' The call to DWYL, she argues, has the effect of encouraging people to believe that the labour they are providing is for their own benefit rather than for the firm that is extracting that labour from them. Once this mindset is in place, it is that much easier to offer low-paid entry-level jobs, or indeed, unpaid internships, on the grounds that, well, you mightn't be paid much but you are doing what you love.

Economist Richard Denniss in a talk he gave at the Women World Changers conference in October 2017 beautifully illustrated the perniciousness of this. Denniss told the story of asking a group of CEOs how they felt that the women they paid to look after their kids in childcare were among the lowest paid workers in the country. One CEO responded, 'I think it's great. I would much rather have someone looking after my kid who was doing it for love rather than for money.' The same logic enables universities themselves to offer work on a sessional basis, attracting well-trained and dedicated workers who are doing what they love, but doing it under conditions of ongoing insecurity. In the 2011 report *The Academic Profession in Transition*, the authors note that 60 per cent of teaching at Australian universities is done by 'sessionals', those employed part-time on short-term contracts, who have few benefits and typically are not even allocated office space. According to research by the American Association of University Professors, 41 per cent of employees at US universities are adjunct professors who receive low pay and few benefits. Tokumitsu argues that few other professions 'fuse the personal identity of their workers so intimately with the work output', and that because 'academic research should be done out of pure love, the actual conditions of and compensation for this labor become afterthoughts, if they are considered at all'.

If we encourage students to find out what they love so that they can study what they love with the ultimate purpose of finding a job that they love, then we also risk embedding a mentality that encourages workers to devalue the work that they ultimately do, to see self-fulfilling work as a reasonable trade-off for insecure, low-paid employment. This might suit those doing the employing, like the members of the BCA, but it hardly helps employees. The problem with telling people to do what they love is not that it is wrong in itself, but that it is tied entirely to the notion of 'getting a job'. The failure of imagination is one that cannot see outside a world of work and allow that the DWYL approach could actually refer to something other than paid employment.

But it doesn't matter what motivation you choose to encourage people to study and skill up, or how practical you are in tying education to getting a job if, at the end of the day, there are no jobs. We don't have to reiterate the arguments covered in the Work chapter about what the future of work actually looks like, but we do have to, at least, accept that work will be different and therefore so will education. So let's avoid the more controversial claims about a postwork future in which machines do next to everything that we currently call a job, and imagine instead the scenario outlined above in which we work alongside machines, perhaps in a deskilled environment, and where the average work week is reduced considerably, to three or four days a week.

It is likely to be a work environment where we change jobs more regularly than workers have in the past, and where the work we do is increasingly on limited-term contracts directed at particular projects. Workplaces themselves are likely to be increasingly flexible and contingent, providing space to bring teams together on an ad hoc basis rather than the traditional office space used by single-function firms today. Such teams are liable to be diverse in the skillsets they bring together and in the people they attract to work in them, and diverse across gender and nationality as well. Depending on the pay and conditions under which such work is done, such a scenario may create a more enjoyable and fulfilling type of work than the traditional model of picking a single job, becoming expert in it and progressing through a hierarchy of work across a lifelong career. But let's focus on what all this means for education. Such a scenario is quite different to the one that underpins the logic of the BCA report, and it leads to different conclusions about what education for the 21st century should look like.

A 2017 report, *The New Work Smarts*, by the Foundation for Young Australians (FYA), takes this sort of flexible, contract-based, ever-evolving career as a given (based on research into the sorts of skills employers are already asking for, extrapolated into the immediate future) and, not surprisingly, conjures a rather different vision for the future of education. Although FYA, too, sees lifelong learning as likely to be the norm, it doesn't see it as being underpinned by the traditional school/higher education/degree-based pathway that BCA favours. The FYA report says that the traditional employment relationship is

> likely to become more fluid with people holding portfolios of activities, including paid employment, unpaid employment (internships or volunteering) and self-employment. Young people will need to be prepared for a journey of lifelong learning and be confident to work autonomously. They need to be critical thinkers and problem solvers, but even more importantly have strong communication skills to interact with people.

This means that jobs cannot be the focus. Rather, they say, 'if we want young people to capitalise on these opportunities and navigate the challenges brought by these changes, they need a set of transferrable skills. We must equip young people with the new work smart skills and capabilities: smart learning, smart thinking and smart doing.' FYA lauds an education system it already sees emerging in some parts of the world, where 'the most progressive education systems are focusing on developing the "new work smart" workforce of the future'. Such countries 'offer immersive, project-based and real-world learning experiences that go beyond the classroom environment, such as working with local businesses or facilitating art and film projects in local communities'. These learning experiences are 'best suited to developing the future-proof enterprising and career management skills that will be most in demand and most highly portable in the future of work, and instil in young people the enthusiasm for ongoing learning that will be critical for their future success'.

In fairness, there are major areas of crossover between what BCA and FYA say about education and the work of the future, though it is interesting that FYA ends up being much more optimistic despite the future it projects being in many ways more uncertain than that projected by BCA. Whereas BCA wants just to 'future-proof' us, FYA is much more willing to embrace a future of work that is rewarding as well as challenging.

But what happens if we push the scenario a little further than either of these reports and into a postwork future? What if we allow for a future where we not only work less, but where fewer people are needed to work at all? Both BCA and FYA avoid confronting this scenario, but it is worth contemplating. Again, I don't want to reiterate the arguments raised in the Work chapter about that 'robot question', but we do need to consider the possibility that a workless – or work less – future is one we could actively choose. If we did, education would have to be radically rethought too, as its key focus would no longer be about making us 'job ready'. Education would be more about learning for a world of not-work.

Let's look at it from a point of view that says we *should* be trying to build a society in which paid employment is less central to how we structure society and our role in it, and where we actively pursue policies that embrace the job-destroying possibilities of new technologies. As I say in *Why the Future is Workless*, a 'postwork future is not one in which people no longer do anything recognisable as what we would today call work. It is one in which we are liberated from the compulsion to work for a wage in order to survive.' Or as British political commentator Aaron Bastani describes it: 'If we embraced work-saving technologies rather than feared them, and organized our society around their potential, it could mean being able to live a good life with a ten-hour working week … Cartier for everyone, Montblanc for the masses and Chloé for all.' An outcome he calls, somewhat tongue-in-cheek, fully automated luxury communism.

The primacy of work, of having a paying job, is so deeply embedded in our societies that the very idea of postwork can sound unconvincing at best, dangerous and ridiculous at worst. It is almost impossible for most of us to imagine what a postwork world would even look like. And yet, we know that a traditional understanding of work is no longer viable for many people. The promise of paid employment – the very thing that makes taking on huge student debt in order to get a degree and therefore a job seem reasonable – is routinely broken as the jobs we do fail to even financially support us, let alone provide the sort of emotional satisfaction and personal fulfilment we are routinely told justifies this lifelong commitment to work, work, work. According to the International Labor Organization (ILO), global unemployment reached over 200 million in 2017, with one in eight of the global youth workforce, more than 70 million young women and men, unemployed. Real wage growth in 2015 averaged 1.7 per cent but was only 0.7 per cent if you leave out China. Since the 1990s the amount of wealth going to wages has declined greatly in many advanced and emerging economies. Gender inequality persists in global labour markets, in respect of opportunities, treatment and outcomes. If you look beyond these figures, which count people in work – and which define 'in work' as one hour of work per week – there is a whole other strata of people the ILO defines as being in 'vulnerable employment', those eking out a bare existence in developing countries. Globally, 760 million women and men are working but not able to lift themselves and their families above the $3.10 a day poverty threshold. In fact, the ILO says, 75 per cent of workers are employed in temporary or contingent jobs or in small family businesses. So even where people have jobs, they can barely make a living. I mean, for heaven's sake, 45 per cent of homeless people in the United States have a job, and two-thirds of those in Britain living below the poverty line are in working households.

Do we honestly think that all this is going to turn around anytime soon, and that all these people are going to suddenly find sustaining, middle-class work? Is sending everyone off to university really going to make things better? James Livingston points out in *The Baffler* ('Why work?'):

> A quarter of the adults employed in the United States are paid wages too low to lift them above the federal poverty line ... Nearly a quarter of American children live in officially defined poverty: food stamps and emergency rooms keep them alive ... every Walmart with three hundred or more 'associates' costs taxpayers roughly a million dollars in public assistance each year because the wages paid these employees don't cover their food and health care ... The requisite jobs don't exist, and almost half of those that do don't pay enough to live on (much less build anyone's character). The United States is a place where hard labor means a prison sentence, not a living wage, and work means economic impoverishment, not moral possibility.

A survey by YouGov in the UK found that '37% of working British adults say their job is not making a meaningful contribution to the world' and that 'only 18% say it is very fulfilling'.

As I've noted, education does improve your chances of getting a job and those with a degree are more likely to have a higher wage. Nonetheless, education is no longer the sure-fire road to job success that it once was, and the job you have is likely to change far more frequently than in the past. I was on a panel with HR specialist Jennifer Brice in July 2017, discussing the changing nature of work, and she pointed out that some of the big firms she deals with, such as IBM, conduct exit interviews with new employees on their first day of work. They simply presume new recruits won't be sticking around. In 1998, 85 per cent of students could expect to have full-time jobs in the area of their degree within four months of graduation. By 2014 that number had dropped to 66 per cent. And naturally, student dissatisfaction with education is increasing: a 2012 report, the *Australasian Survey of Student Engagement*, showed that only 37 per cent of students thought their degree had developed the skills they needed for a job, and only 27 per cent thought it had prepared them to work with others. Writing in *New Matilda* ('University paradoxes'), Cat Moir notes a report released in early 2018 by Quality Indicators for Teaching and Learning called the *Employer Satisfaction Survey*, which suggests that 'up to 38% of graduates leaving Australian universities today will not find full-time work' and that 'the last decade has seen a rise of 17% in the number of university leavers in part-time employment'.

Whether we are imagining a postwork future in which machines do much of the work now undertaken by humans, or simply one in which technology creates new work for humans, we are going to need an education system that doesn't educate for jobs per se, but that educates for learning. Under such circumstances, the sort of skills we will need are more likely to be the so-called soft skills of human interaction, empathy, problem solving and creativity rather than specific technical skills. Indeed, simply concentrating on technical skills could end up being nothing more than a training in obsolescence, as technologies rapidly change and AI substitutes for many of the tasks that make up the jobs that we do. These humanities skills and disciplines tend to be overlooked partly because they do not submit to the sort of quantitative analysis and measurement preferred by neoliberal society, such as productivity, so it is hard to measure their 'outputs'. To a large extent, this makes them invisible, or irrelevant, to the neoliberal state and they are not so much undervalued as not valued in the first place. Despite this, the changing nature of work, and dissatisfaction with educational outcomes, is forcing a reassessment.

Peter Acton, president of Humanities 21, a not-for-profit advocacy group that lobbies governments and businesses to take humanities training more seriously, writes that employers

> are looking for people who are not just able to draw on established practice that delivers proven solutions, but who know how to deal with ambiguity, work with inexact or incomplete information, explore different possibilities and judge how to advance a conclusion or course of action where there is no proven approach or answer.

He notes that this 'is reflected in job advertisements' and that 'company recruiters are now more likely to specify enterprise skills than technical skills, and over the last three years the proportion that demand critical thinking has increased by 158 per cent, creativity by 65 per cent, presentation skills by 25 per cent and teamwork by 19 per cent'. These capabilities are 'not about learning the prescription to achieve a textbook result [but] about having the intellectual capacity to attack those issues for which there is as yet no metaphorical text or answer'.

This flies in the face of the oft-heard calls for children to be taught coding at school or for university students to do STEM subjects, but the reality is, the call for such specialist training is greatly overstated. At the very most, it is the T-subjects, technology, which might be in demand, not the entire acronym. Michael S Teitelbaum, an expert on science education and policy, told *The New York Times* that 'executives and lobbyists for technology companies do a disservice when they raise the alarm that America is facing a worrying shortfall of STEM workers'. He says, 'We're misleading a lot of young people.' This is borne out by a recent report released by Google that looked at which employees within their company were most successful. Google was founded on the principle that only those with technology training can understand technology, but the findings from Project Oxygen, as it was called, contradicted this confident assertion. *The Washington Post* reports that 'among the eight most important qualities of Google's top employees, STEM expertise comes in dead last'. Project Oxygen found that the

> top characteristics of success at Google are all soft skills: being a good coach; communicating and listening well; possessing insights into others (including others' different values and points of view); having empathy toward and being supportive of one's colleagues; being a good critical thinker and problem

> solver; and being able to make connections across complex ideas.

Google has adjusted its hiring policies appropriately, including in the mix of potential employees those who have standard arts degrees.

All this means that the last thing an education system based on technical skills is likely to do is to future-proof us. You are more likely to be future-proofed by getting a set of transferable skills than you are in thinking of education as primarily training for a specific job. What's more, the beauty of a less job-focussed approach to education is that it is more likely to prepare us not only for the future world of work, but also for a world in which we all work less, or even work not at all. Education needs to return to the principle of preparing good citizens, not just work-ready employees, and the upside is that an education in these soft skills prepares us for both. Peter Acton summarises the situation well when he says:

> We are increasingly called on to embrace [wicked problems that have no precedent, such as climate change and how to implement an ethical approach to managing artificial intelligence] and other unknowns. We cannot apply past answers because the past gives no direct answers (though it may give plenty of clues). Our only option is to discover or invent the solution.

We therefore need to do more than 'just prepare young people to survive in the world of unknowns; we need to equip them to manage the unknowns and thrive'. And really, the best way to do this is by embracing education focussed on the humanities. 'In an increasingly secular society', Acton says, 'the liberal arts provide a critical set of disciplines that look both inwards and outwards, and draw on the special qualities of human experience and intellect to help us navigate it all.'

The education system we inhabit is beyond its use-by date, 'an anachronistic relic of the industrial past', as Jennifer Gidley calls it. She says it fragments and compartmentalises knowledge; it advances science over literature, mathematics over art, intellectual over emotional qualities, materialistic over spiritual values, and the status quo over creativity. It also privileges neoliberal managerialist practices, which in turn encourages neoconservative research agendas 'linked to an audit culture' instead of 'qualitative, creative inquiry'; and educates for an outmoded way of thinking unsuited to the complex, networked, global environment in which we now live and work.

Responding to these shortcomings, Silicon Valley industrialists and other tech types, unsurprisingly, want to double down on the sort of thinking that works for them in their approach to business, and want to 'disrupt' education. This involves 'unbundling' education services, which means creating 'new providers' who use technology to offer shorter, modular courses, focussed entirely on targeted jobs training and offered at a lower cost than courses at traditional institutions. This is the classic 'disruptor' approach, finding a small market somewhat neglected by the big operators in the industry, into which a new operator can move and eventually undermine the main players. As I have suggested above, such modularity may be a useful approach in some circumstances, but it is really a business methodology, not a pedagogical one. All this approach does is reinforce the worst aspects of the current system – most especially the relentless focus on jobs and marketisation – while treating technology as a *deus ex machina* rather than what it really is: a tool that can be deployed strategically. It would be ridiculous not to take advantage of the immense amount of knowledge and information a smartphone or other device connected to the internet puts at our fingertips, but we have to integrate that into teaching in a creative way, not use it as a substitute for teaching. A report by conservative Washington think tank the American Enterprise Institute sets out an argument for the unbundled approach, but you don't get more than a few pages into it before you notice how oriented it is towards education as a product, the very neoliberal model and logic that we need to move past. The report talks in terms of a 'chartering model' that 'would empower new, independent authorizers to govern market entry and hold providers accountable over time' and 'reformers' developing an 'outcomes-based approach focused on value that measures labor market outcomes and student satisfaction relative to an institution's total expenditures'. It argues, 'policymakers might also choose to wait for the market to mature on its own and let consumer demand and competition drive innovation'. 'Market pressure', it says, 'not government, may better facilitate the emergence of a high-quality unbundled market.'

When we treat education as a product, we create customers, not students, let alone citizens. We train rather than teach and risk locking ourselves into a model that might prepare people for a job – provided there are any – but not for anything else. It is worth noting that the same people who want to 'unbundle' education are doing the same thing to jobs themselves, with the result that tasks can either be outsourced to overseas workforces at a fraction of the cost or be replaced by machines altogether. They are advocating an education system to prepare us for the jobs that they themselves are destroying. This moves beyond irony and into farce.

If we really want education to be something other than a product, to be a way of preparing for an uncertain future of immense possibility, with or without a job, then modularity and technology are only ever going to be a small part of the solution. The self-defeating nature of the 'unbundled' school should be obvious: if the world of work is increasingly about preparing for people to do the things that machines can't, how can we possibly envisage a system of education that is anything other than grounded in humanness? If the skills that are valuable – either in work or in life – are the soft skills of human interaction, everything from creativity to empathy, how can we possibly rely on modular, market-driven teaching conducted by machines or online? For sure, we can use those tools as part of the learning process and, especially for postgraduate students, the ability to use a MOOC to acquire a skill they don't have can be invaluable, but as the focus of education, it is bound to be a failure.

Education needs to be taught by professionals who have the time and security to properly engage with the students in their care, not sessionals struggling to make ends meet, probably holding down several other gig jobs. It needs to be taught in a cross-disciplinary way, not in silos where the engineer is walled off from the sociologist, the physicist from the artist. To break down those walls, there needs to be less reliance on lectures, note-taking, testing and all-round regurgitation, and a shift to the application of a range of knowledges taught in problem-solving situations. Education has to be as freely available as possible so that the institutions of education are as diverse in areas of gender, ethnicity and class as possible. The message we should be sending to children, teenagers and young adults at every stage of their education is not to do what you love, with all the hidden horrors that advice entails, but to simply learn for love. As Gidley says:

> If young people are to thrive in educational settings, new spaces need to be opened up for softer terms, such as love, nurture, respect, wonder … openness and trust … Focusing on love as the heart of education counter-balances the technologies and commodification of education and resists the censorship of the audit [neoliberal] culture.

To treat education in purely instrumental terms, as a means to the end of a job, is to foreclose on a future we cannot see. Worse, it is to drive us in a direction we do not need to head, limiting any sense that another world is possible. If the realm of education, of all things, is reduced to this single function, how will we ever aspire to more? How will we devise better ways of living if we simply presume that the only acceptable life is to have a job defined by another era? Seriously, is the limit of our ambitions as humans that we will be cogs in the machine, that we will forever and ever be an employee or an employer? Education should be about possibilities, even remote ones, not about future-proofing and channelling us into a predetermined future where it is impossible for us to imagine something better. Something different. Something *else*.

Under such circumstances, the prospect of 'lifelong

learning' starts to look like another treadmill, another production line, a gauntlet we run, not a pleasure we embrace. We need to recognise that the best jobs training is also the best life training, a philosophical approach that favours learning above the acquisition of mere skills.

And this brings me to joy.

JOY

> Power without love is reckless and abusive, and love without power is sentimental and anemic. Power at its best is love implementing the demands of justice, and justice at its best is power correcting everything that stands against love.
>
> *Martin Luther King*

Much of this book is concerned with inequality and political power, with the institutions of the state and civil society, with how we might better govern ourselves to have a better life. I have talked about work and wealth, government, media and education, and have suggested ways of changing how we think about all of them. The unifying thread is a life in common, finding ways to give ordinary people more power over their day-to-day lives, whether by putting them at the heart of our major institutions or by providing them with the material means to participate in society. In a complex and necessarily interdependent world, we need to build institutions and systems that allow us all to thrive on our own terms, not ones that force us into making a tiny elite richer than they already are. The technicalities of how we design these institutions and systems matter, and

I hope I have given them adequate attention – including a theory of power that might allow us to achieve them – but these are merely means to ends, and it is those ends I will talk about in this final chapter. In so doing, I will push the idea of 'a life in common' out into deeper water. While it's important to speak about economic equality and to fight for a world in which everyone shares fairly in the wealth we all help create, we can't equate happiness and success solely with material well-being. I don't want to play down the importance of material satisfaction in constructing a decent life, but I do want to stop pitching all our endeavours at that alone. And so, here at the end of the book, I'll bring together some ideas that culminate in what I will call a theory of joy. This is not just about pursuing change that makes us happy – though it is that – but about recognising that the pursuit itself can be joyous. It is about the shared joy of building a life in common.

I have suggested that we need big, audacious reforms for a reason. Such thinking snaps us out of the complacency of the status quo and helps break the habit of thinking that 'there is no alternative'. We need ambitious ideas for change and the courage to pursue them, but to build that courage, we need to fiddle around the edges too. We need to bring the spirit of *tinkering* to democratic governance. Tinkering, the practice of do-it-yourself (DIY), or make-it-yourself (MIY), or do it-together (DIT), is a growing response to the disposable, closed-system consumerism that so dominates late capitalism. Instead of throwing stuff out and buying new things, tinkerers want to keep them and fix them, and maybe we could apply these values more generally. In her wonderful book *Tinkering*, Katherine Wilson says, 'committed tinkerers have much to teach all of us about negotiating power and living well, especially amidst the current conditions of job insecurity, wage disparity, career precarity, globalisation, privatisation, housing unaffordability and deskilling in all sectors – and the disintegrative impacts these can have on our livelihoods, identities and humanity'. To upend the mindset of disposability and consumption-for-consumption's sake, we need to realise that some things can be fixed rather than thrown out, and that in fixing them, we can sometimes transform them.

Tinkering, Wilson says, is a vision

> of freedom, of pride, dignity ethics and artisanal joy – an unfettered way to live according to one's own measures. It can be practical and utilitarian … but also a form of scholarship, play, adventure, resourcefulness and resilience. It can be a portal to social connection, community, spirituality, thrift, identity and political resistance.

Perhaps such an approach can encourage us to be much braver about our ability to muck around with our democratic system, or invent workarounds, or mend what is broken, or reroute around blockages. This is the essence of the grassroots organising I mentioned in the Power chapter; it is the essence of finding local ways to manage any given commons. We can hold meetings, public talks and fundraisers, and we can get things done if just have confidence enough to trust ourselves over those who tell us we are disengaged or lack the expertise necessary to participate. As I have illustrated with the examples of citizens' juries and deliberative polls (in the Government chapter), ordinary citizens have a lot to offer – and are willing to offer it

– when the institutional frameworks exist that allow them to be involved. The tinkering mindset, of starting small and fixing what we can, may be the way for us to build the ambition we need to tackle the more audacious reforms.

So we need to be ambitious, but ambition, like so much else, has been privatised. When we think of visionaries, of those with big ideas, with the ability to rally people to get things done, the people that tend to come to mind are entrepreneurs, the tech giants and wizards of Silicon Valley, the masters of the financial universe, the thrusters and chancers of the commercial world. We have almost become embarrassed by the idea of the public intellectual, the public philosopher or the sage individual, or even the spiritual leader who can show us a better way. These categories, too, have been privatised, and we are greeted with the commodified outpourings of book tours, TED Talks and mindless shoutfests on television programs like *Q&A*. Neoliberalism encourages the creation of individuals as 'brands' pontificating from on high, rather than collective deliberation rising from below. It is harder than ever to elbow these attention-grabbers out of the way, to clear the table, so that we can do a bit of tinkering ourselves.

Where we imagine social policy at all, we tend to see it as a matter for government, for political parties, and we outsource our concerns to them, which has led to the dissatisfaction that affects most democratic nations today. We have lost our sense of control over these things. We see those policies and those governments as increasingly at the beck and call of the wealthy and well connected, and no longer open to, or caring about, the rest of us. Still, there are reasons to be hopeful, evidence that we are increasingly willing to tinker with the system, to eke back control, even as it militates against our involvement. In Australia, for instance, we are using our tools of democracy to alter long-established patterns of voting. Where once most of the vote was divided between the two major parties, they now struggle to get 30 per cent of the vote each. The Greens are a workaround for the traditional two-party system. Pauline Hanson's One Nation Party is a patch that some citizens thought we needed. Independents such as Cathy McGowan and Andrew Wilkie are a DIY fix by the people in their electorates. And the crossbenches are growing. In towns as far apart as Cleveland, Ohio, and Preston in the United Kingdom, residents are reinventing the way their local governments operate, rebuilding communities that had all but been abandoned by central government. The so-called 'Cleveland model' is using workers' cooperatives as way of rebuilding towns left behind by globalisation.

'We must have legends or we will die of strangeness,' the poet Les Murray once told us. We need those legendary individuals who can focus our attention and look past the near horizon to show us that hope is possible. These leaders are important, but they need to emerge, not be imposed – and how wonderful it has been to see the young people most affected by the insanity of America's almost non-existent gun laws take the lead where their politicians have so obviously failed. The likes of Delaney Tarr, Emma González and David Hogg emerged when the sensible centre had given up entirely.

Ambition has been privatised, which means it has also been commercialised, and it is about time we became ambitious for a life that means more than buying stuff. Don't get me wrong, I am someone who likes his iPhone, who is pretty happy to spend time on social media and who uses

Uber. I get the attraction of owning a really big house and a top-of-the-range Tesla, though I don't really want either. For heaven's sake, I have a Nespresso coffee machine (don't @ me). But our ambition needs to extend to something more than consumption and convenience, either as an end in itself or as compensation for the other absences in our lives. We have to be able to step back from the happiness of immediate gratification and instead embrace the joy of living, even when living isn't particularly joyful.

Richard Denniss tackles this idea in his book *Curing Affluenza*:

> We have built a culture where buying things is increasingly unrelated to using things. And we have built a culture where things are thrown away not because they are broken, but because they send the wrong signal about who we are ... In turn we have built the most materially wealthy communities the world has ever known, but despite this abundance of stuff, our culture makes people feel that they never have enough, or the right, stuff.

His key point is that curing 'affluenza', as he calls our tendency to over-consume, means that 'we will waste far less time and far fewer resources, and in turn make far more of the things we really want more of' if we are more thoughtful about the way we consume stuff. We will, he suggests, extract more pleasure from the things we own than we do now, simply by exercising a bit of judicious discrimination. Denniss is not concerned with us becoming less materialistic per se, nor is he advocating we abandon the material world, put on a hairshirt, and go and live on berries in the forest. What he *is* saying is that we need to think differently about the way we approach our economy, and his central point is that it is the *shape* of the economy that matters and not its overall size. In an article in the *Sydney Morning Herald*, he notes that for decades we have been told 'as long as the amount of stuff getting bought is growing, we must be doing well'. But while 'there is no doubt that opening new asbestos mines would lead to an increase in GDP in Australia, there is no evidence that selling more of something that harms us would make us "better off"'. Growth alone doesn't necessarily improve our lives – growth in some areas can actually make us worse off – but despite this, Denniss says:

> Just think about how often you have heard a politician or a business leader say we need to 'grow the economy' without saying which parts of the economy they want to grow. Indeed, think about the number of times you have heard a politician say we need to shrink parts of the economy that people want more of, like the health sector and the education sector, to grow the economy overall. What a strange idea.

Another way people approach the affluenza problem is to distinguish between needs and wants. We should, the argument goes, diminish our wants and concentrate instead on satisfying our needs. But this distinction isn't as straightforward as such people presume, as I can illustrate with an example. For dinner last night we had a simple but delicious meal of roast lamb, hummus, tabouli and flatbread. It's a family staple. But I was struck by how much the experience was enhanced by the plates the meal was served on, beautiful bowls and dishes we have been given as gifts or have come across in our travels. I cut the meat with a hand-crafted Japanese knife, sharpened the way that a man who had dedicated his whole life to perfecting the art of blade sharpening had showed me. The sheer pleasure of preparing the food with such an implement, and serving it on these vessels, conjured a feeling of happiness that goes beyond any utilitarian benefit that comes from the food alone. The vessels and the tools provided a level of elevation, and it is this elevation that blurs the line between needs and wants.

Historian Humphrey McQueen notes in his book *The Essence of Capitalism*, 'Needs cannot be categorised as true or false, biologically determined or socially conditioned. To convict marketeers of stimulating false wants, the prosecution must first identify true needs.' This is impossible to do in any systematic way, he suggests, and 'to label the minimum that we require true and then disparage every need over and above the rawest necessities as false is to deny humanity our capacity to remake ourselves. Civilisation exists because we have spread our needs beyond those essential for existence.' He quotes an evocative passage from Marx: 'Hunger is hunger, but the hunger gratified by cooked meat eaten with a knife and fork is a different hunger from that which bolts down raw meat with the aid of hand, nail and tooth. Production thus produces not only the object but also the manner of consumption.' McQueen continues, 'A compulsion to drink cannot explain the decorations on vessels or the rituals around a bar ... To survive longer than a year, hunter-gatherers must provide themselves with tools and shelter. To thrive across generations, they require weapons and territories, language and song, affection and companionship, magic and ceremonies.' In the absence of these things we become disenchanted, literally without magic. McQueen's book is about the history of that ultimate capitalist product Coca-Cola, and he notes that it came into existence at a time when real wages were rising and people therefore had the money to spend on more than the basics of food, shelter and clothing. He says,

> liquids are the essence of life [and] we die of thirst long before we die of hunger. The need to drink water is fixed in our physiology [but] physiology can never explain why billions have paid for a Coke when most of them can get water from a tap practically for free.

None of this is to suggest that the unlimited consumption demanded by capitalism is somehow natural, far from it; but nor is it simply a matter of chastising people for their 'bad' decisions in allegedly satisfying wants at the expense of needs. Citing work by Canadian political economist Michael Lebowitz, McQueen replaces the distinction between needs and wants with the idea of capital-induced needs, and thus brings us back to the role of work in the underlying economic and social system of neoliberalism: 'Capital multiplies our needs, but in ways that further subjugate our lives to its need to expand. The result is that "each new need becomes a new requirement to work".'

It is impossible to choose perfectly between what we need and what we want, what we fight for and what we ignore, or to choose between all the other incommensurable ends that are the natural state of the world. The really tough choices are never between the good and the bad thing – that choice is easy – but between two good things, say, free speech and privacy, or the natural environment and food production. Sometimes we simply can't choose, or we can't choose simply, and we need a way of navigating our choices. Partly this is about political decisions, and most of the book has been dedicated to them. But politics is not enough. We need to give some thought to how to be in the world, with all its inherent problems, inevitable disappointments and its primary characteristic of not owing us anything, and still have a fulfilling life. To do this, we must recognise that we are of the world, even those parts of it we don't like, and therefore our ability to live a good life requires us to deal with the ups as well as the downs, with failure and loss as much as success and happiness. Our theory of joy, then, needs to encompass life in all its complexity and allow us to come out of any situation – good or bad, happy or sad – with our sense of ourselves and our place in the world intact. It needs to encompass the fact that the joy we experience rarely arises in isolation from others, and that it mostly occurs in the context of challenge and difficulty rather than ease and comfort.

At any given moment, we have the choice of whether to build a life in common or to hollow out some space in which we alone, or those like us, can escape the demands of community, of society. Indeed, our cultures are full of stories and mythologies of escape, of starting again, starting over, this time on our own terms and with our own kind. Australia itself, as a Western nation, was founded on the idea that another Western nation, England, could make its own land more pure by dumping its riff-raff on the other side of the world. That this prison colony grew into one of the most successful democracies on earth is a rather wonderful thing, but in the early years of Australia's independence from Britain, a similar desire for 'purity' manifested in the new nation passing the *Immigration Restriction Act*, designed to stop non-Europeans settling in the new land.

America, too, was colonised by Europeans seeking escape and the chance of a new world, and its subsequent history is full of the discourse of independence, individualism and severance from the mainstream. These wishes take forms as various as *Little Orphan Annie* comics and the Bitcoin cryptocurrency. Harold Gray wrote the comic strip *Little Orphan Annie*, which imagined a free market utopia created and run by Daddy Warbucks, and its themes were often critical of organised labour, Roosevelt's New Deal and, of course, communism. Similarly, Bitcoin is what author David Golumbia calls 'software as rightwing extremism'. He sees it as part of a broader 'cyberlibertarian' movement that has deep roots in the right-wing US intellectual history of separatism. Quoting academic Langdon Winner, Golumbia describes it as 'a belief system [that] "links ecstatic enthusiasm for electronically mediated forms of living with radical, right-wing libertarian ideas about the proper definition of freedom, social life, economics, and politics"'. It is a tool – and a philosophy – for escaping the rule of central banks and other statist impositions. To the list of such schemes we could add the current plans of various tech entrepreneurs to colonise Mars and, most importantly, the ideas and novels of Ayn Rand and her whole notion of 'going Galt'. The basic thrust of such movements is to make capitalism safe from democracy.

Of course, you don't have to delve into the depths of libertarianism and other right-wing movements to find fantasies of escape and isolation, of separation from the general mass of society, of fencing yourself off from the world around you, either spiritually or physically and often both.

The Hippie Trail of Southeast Asia, taken by those seeking enlightenment or fulfilment, or the communes of various religious and semi-religious sects, as well as communities built around theories of pedagogy such as Steiner and Montessori, are all part of the same basic urge. And nor should we characterise all such desires as weird or wrong, or conclude that they are driven by greed and anti-democratic urges: they are perfectly understandable and speak to a real need we all feel to lift the burden of social living from our shoulders. Because it is a burden, we need not deny that. Whether it is our obligations as a member of a family, of a community, or of a polity, a life in common lived with other people is always going to make demands on us that will have us entertaining fantasies of escape and seclusion, and some of us may even pursue them.

And yet.

Such escape can never be available to the vast majority of people, and so we simply have to find ways to live together. Democracy is probably the most successful – or least worst – system we have come up with to achieve this, and within it, more specific philosophies – in particular, socialism and liberalism in their many guises – coexist and try to guide us. But with this theory of joy, I am stepping back from these familiar and endlessly contested theories and philosophies to say something instead about how we might simply see ourselves as members of the various communities we inhabit. I have built this book on the idea of a life in common, of participation, by those of us who are so inclined, in running as much of our lives as possible, whether in the workplace or in government itself. I suggest that this participation is not just a means to an end, a grind that we have to go through to reach our goal, but that participation is worthwhile in and of itself. And not just worthwhile, but a source of joy.

The term the philosopher Hannah Arendt uses for this concept is 'public happiness'. She suggests that this is what the American founding fathers meant when they wrote into the Declaration of Independence the inalienable right of the 'pursuit of happiness'. So not private happiness, she suggests, but the happiness that arises from public participation, of being an actor in the administration of your world. This deep pleasure, joy, of participation was understood and built into the invention of democracy, which was why the citizens of Ancient Greece (as circumscribed as that category was) used sortition and not voting to choose their government: the ruled always need the opportunity to be the ruler. We began to lose this public happiness when we made our democracies representative, handed them over to an elected elite, and then contented ourselves almost exclusively with the private pleasures of home and hearth. This was duly exploited by capitalism, which has a vested interest in creating and satisfying needs and wants of the private rather than the public sort, and so gradually we lost track of the very idea of public happiness. It is this public aspect that our theory of joy needs to rediscover because, in Arendt's words, 'No one could be called happy without his share of public happiness ... no one could be called free without his experience in public freedom, and ... no one could be called happy or free without participating, and having a share, in public power.'

Oscar Wilde's quip that the 'trouble with socialism is that it takes up too many evenings' no doubt resonates with many – as does Max Weber's description of politics as 'the strong and slow boring of hard boards' – but we shouldn't let such insights turn us completely inward. A share in public power, control over our own lives, is an end in itself and a joyous one.

Another writer to explore this idea is Lynne Segal in her book *Radical Happiness*. She draws on Arendt and many others, but most importantly, her own experience as an activist feminist, to articulate a view that, since 'our own survival and identity depend upon the existence of others and the communication between us, neither our emotions nor our deeds are ever purely private affairs'. She stakes out the deep claims of collective joy against the ephemerality of the individual version, not to dismiss individual joy but to enhance it. Radical happiness, as an aspect of a life in common and the practice of governing ourselves, is at once surprising and transcendent:

> From its religious origins, through to the Romantics' love of nature, in moments of shared political passion or in what remains today of public festivals, joy is most often associated with experiences that take us altogether outside ourselves. They are either intense feelings of pleasure that we share with others, or that others, at least potentially, might share. I think this also tells us something about personal happiness. As in drug-induced euphoria, what we recall later as pleasurable often occurred when we were fully absorbed in our experiences of the moment, whatever the context, whether reading a book, gardening or arguing passionately with friends. In these moments, we do not worry about whether we are happy or not, any more than a child does when fully engrossed in life at large.

Speaking personally, I am your stock-standard fallible progressive who doesn't really live up to his best ideals, is trying to do better, but who recognises his own limits. I am all for live and let live, and staying out of your way and letting you get on with things, whatever your thing is, but I also recognise that we are all in this together and that most of us, left to our own devices, are likely to fuck it up for everybody else. The more power we get as an individual, the more likely we are to think that we deserve it, that we are self-made men and women, and this will ultimately make it easier for us to ignore everyone else and rationalise away their pain. A life in common *uninsulates* us from the experiences of others and therefore saves us from ourselves. Paradoxically, it empowers us, but it disperses power so that it is not centralised into something despotic.

So this book is me coming to terms with the problems of the world as I see them, within the confines of the human failings we all share, and mostly it feels like trying to juggle a kitten and a chainsaw at the same time, with a bowling ball thrown in for good measure. It is a hugely arrogant undertaking for anyone to offer their views on the future of everything, so it's important to recognise how partial the discussion here is. I can rationalise what I have left out by saying, first, you can't cover everything: the title is a conceit, a trick of the eye designed to get you thinking in big-picture terms. If I'd called the book *Six Ideas for a Better World*, or something along those lines, you would read it with much more restricted expectations. So whatever I have failed to supply to fit the scope implied by the title, I encourage you to supply yourself. Fill in the gaps.

More importantly, the ideas I offer are predicated on a sense of human universalism, the idea that we are all in this together, that there is no alternative to our shared existence, and so therefore it is best – wise, even – to work from the premise that we have more in common than we do in opposition. This, in turn, is based on the idea that we have only one planet and that its environment is necessary to our ongoing existence, no matter what tiny part of the planet we live on and no matter what type of human we are. We all live or die with the health of the planet. Universalism, then, is not just an airy-fairy incantation chanted in an incense-filled room while someone plays the sitar and tambourine: it is a rock-solid practical presupposition that none of us can prosper without.

Still, at any given moment, this universalism is challenged and made shabby by the lived experience of particular lives. Whether as women, people of colour, Muslims, people with disabilities – or whatever identity we inhabit within the specifics of our universalism – we must recognise the soul-destroying difficulty of what it means to put aside that lived experience in the name of some theoretical universalism. I know that, and I apologise for the extent to which I ride over difference, but universalism is the rock I am standing on, for better or worse. Nonetheless, the life in common I am arguing for here is a universalism that honours differences of race, culture, gender, ability and ultimately individuality. I have no interest in communalism and homogeneity, in assimilation and monoculture. It is multi all the way down.

Raymond Williams once said that it is easier to imagine the end of the world than the end of capitalism, but I don't think that is right anymore. The end of capitalism is all around us, as the planet heats, as the oceans fill with plastic, as our societies lose their coherence, as species die and habitats fail, as economies become ends in themselves. Capitalism hasn't ended but its end is all around us and soon it will be gone altogether. If you want to see the end of capitalism, you can. I am hardly the first to notice this, but it is still something we only speak of in hushed tones and with a little incredulity. If you say it out loud, it still invites jeering and laughter and the smug looks of those who know better, but there it is. Part of the difficulty is that there will be no final moment, like the fall of the Berlin Wall, to mark now and then, before and after. There will just be a long, slow melt into joylessness unless we recognise the opportunity and make something better. There is still time.

The essence of power in a democratic nation is the building of institutions. Once established, such institutions are themselves powerful: they resist attempts to steer us towards tyranny if they embody and practise non-tyrannical values. But we have to build them in the first place, and then maintain them, and that means fighting against those who seek to undermine them. Powerful forces are working to undermine public services everywhere and at every opportunity, and they are not even trying to hide it anymore. They live in a world in which the colour of a person's skin offends them, in which woman are inferior, in which their religion is a guiding light but another's is certain to doom its followers to eternal damnation; and they have atheist counterparts who are just as small-minded. There are people in the world who see nothing wrong with abusing the weak and salivating over the powerful, whether on social media or in less virtual spaces. These people cannot be reasoned with, only defeated.

To do that, the rest of us must rediscover the joy of exercising real political power.

The Japanese have the beautiful practice of *kintsugi* that literally means 'repair with gold'. When a vase or a bowl breaks, instead of tossing it out and buying a new one, they use gold to join the pieces back together, turning the repaired item into a new and beautiful piece, not despite its fractures but because of them. In the same way, we need to rethink our relationship with the institutions of democracy and reclaim our control over them. We are the gold that can make them whole again.

SELECT READING

A*hler, Douglas J & David E Broockman, 'How ideological moderation* conceals support for immoderate policies: a new perspective on *the "disconnect" in A*merican politics', 23 September 2014, <citeseerx.ist.psu.edu/viewdoc/download*?doi=10.1.1.688.8618&rep=rep1&type=pdf*>.

*Allcott, Hunt & M*atthew Gentzkow, 'Social media and fake news in the 2016 e*lection', National Bureau of Economi*c Research, working paper no. 23089, June 2017, <nber.org/papers/w23089>.

Allen, Jonathan P, Technology and Inequality: Concentrated wealth in a digital w*orld, Palgrave Macmillan, Cham, Switzerland, 2017.*

Arendt, Hannah, Crises of the Republic, Harcourt Brace Jovanovich, New York, 1972.

Barnett, Anthony & Peter Carty, The Athenian Option: Radical reform for the House of Lords, Imprint Academic, Exeter, UK, 2008.

Beveridge, William, Soc*ial Insurance and Allie*d Services, HMSO, London, 1942, <www.scielosp.org/article/ssm/content/raw/?resou*rce_ssm_path=/media/as*sets/bwho/v78n6/v78n6a17.pdf>.

Business Council of Australia, Future-Proof: Protecting Australians through education and *skills, BCA, 2017, <bca.com.au/publications/future-proof-protecting-australians-th*rough-education-and-skills>.

Cegłowski, Maciej, 'Notes from an emergency', <idlewords.com/talks/notes_from_an_emergency.htm>.

Chadwick, *Andrew, The Hybrid Media System, Oxford University Press, 2nd edn, 2017.*

Chancel, Lucas (general coordinator), World Inequality Report, World Inequality Lab, 2018, <wir2018.wid.world/files/download/wir2018-full-report-english.pdf>.

Community-Wealth.Org, The Cleveland Model – How the evergreen cooperatives are building community wealth, <community-wealth.org/content/cleveland-model-how-evergreen-cooperatives-are-building-community-wealth>.

Croggon, Alison, 'On power', Overland Magazine, vol. 229, summer 2017, <overland.org.au/previous-issues/issue-229/column-alison-croggon/>.

Dahl, Robert A, *Democracy and Its Critics*, Yale University Press, New Haven, CT, 1991.

Davies, William, 'How statistics lost their power', *Guardian*, 19 January 2017, <theguardian.com/politics/2017/jan/19/crisis-of-statistics-big-data-democracy>

Denniss, Richard, *Curing Affluenza: How to buy less stuff and save the world*, Black Inc., Carlton, Vic., 2017.

Dunlop, Tim, *Why the Future is Workless*, NewSouth Books, Sydney, 2016.

—— *The New Front Page: New media and the rise of the audience*, Scribe, Brunswick, Vic., 2013.

Fioramonti, Lorenzo, *The World After GDP: Politics, business and society in the post growth era*, Polity Press, Cambridge, UK, 2017.

Foundation for Young Australians, *The New Work Smarts: Thriving in the New Work Order*, FYA, 2017, <fya.org.au/wp-content/uploads/2017/07/FYA_TheNewWorkSmarts_July2017.pdf>.

Frank, Thomas, 'To the precinct station', *The Baffler*, no. 21, November 2012, <thebaffler.com/salvos/to-the-precinct-station>.

Friedman, Milton, *Free to Choose: A personal statement*, Harcourt Brace Jovanovich, New York, 1979.

Gidley, Jennifer M, *Postformal Education: A philosophy for complex futures*, Springer International Publishing, Basel, Switzerland, 2016.

Gruen, Nicholas, 'Detox democracy through representation by random selection', *The Mandarin*, 14 February 2017, <themandarin.com.au/75323-nicholas-gruen-detoxing-democracy/>.

Haider, Shuja, 'Safety pins and swastikas', *Jacobin*, 5 January 2017, <jacobinmag.com/2017/01/safety-pin-box-richard-spencer-neo-nazis-alt-right-identity-politics/>.

Harvey, David, *Rebel Cities: From the right to the city to the urban revolution*, Verso, London, 2012.

Heimans, Jeremy & Henry Timms, *New Power: How power works in our hyperconnected world – and how to make it work for you*, Pan Macmillan Australia, Sydney, 2018.

Hess, Charlotte & Elinor Ostrom (eds), *Understanding Knowledge as a Commons: From theory to practice*, MIT Press, Cambridge, MA, 2011.

Ibarra, Imanol Arrieta, Leonard Goff, Diego Jiménez Hernández, Jaron Lanier & E Glen Weyl, 'Should we treat data as labor? Moving beyond "free"', *American Economic Association Papers & Proceedings*, vol. 1, no. 1, forthcoming (27 December 2017), <papers.ssrn.com/sol3/papers.cfm?abstract_id=3093683>.

Klein, Naomi, *No Is Not Enough: Defeating the new shock politics*, Penguin, London, 2017.

Lanier, Jaron, *Who Owns the Future?* Simon and Schuster, New York, 2014.

Luders, Joseph, *The Civil Rights Movement and the Logic of Social Change*, Cambridge University Press, Cambridge, NY, 2010.

Lyons, Tim, 'Is a collective future still possible?' 27 April 2017, <percapita.org.au/wp-content/uploads/2017/05/Fabians-Speech-April-2017-LYONS1.pdf>.

—— 'Solidarity and self-interest in the future of unionism', 13 November 2016, <eurekastreet.com.au/article.aspx?aeid=50237>.

McAlevey, Jane F, *No Shortcuts: Organizing for power in the new gilded age*, Oxford University Press, 2016.

McQueen, Humphrey, *The Essence of Capitalism: The origins of our future*, Black Rose Books, Montreal, Canada, 2003.

Malchik, Antonia, 'Who owns the earth?' *Aeon*, 12 April 2016, <aeon.co/essays/is-it-time-to-upend-the-idea-that-land-is-private-property>.

Manin, Bernard, *The Principles of Representative Government*, Cambridge University Press, Cambridge, NY, 1997.

Mitchell, Timothy, *The Rule of Experts: Egypt, Techno-Politics, Modernity*, University of California Press, 2001.

Mueller, Gavin, 'No alternative', Real Life, 22 January 2018, <reallifemag.com/no-alternative/>.

OECD, 'The risk of automation for jobs in OECD countries', *OECD Social, Employment and Migration Working Papers*, no. 189, 2016, <oecd-ilibrary.org/social-issues-migration-health/the-risk-of-automation-for-jobs-in-oecd-countries_5jlz9h56dvq7-en>.

Orwell, George, *The Lion and the Unicorn: Socialism and the English genius*, Penguin, London, 1990 (1941).

Ostrom, Elinor, *Governing the Commons: The evolution of institutions for collective action*, Cambridge University Press, Cambridge, NY, 1990.

Quiggin, John, 'Doing more with less: the economic lesson of Peak Paper', *Aeon*, 12 February 2016, <aeon.co/ideas/doing-more-with-less-the-economic-lesson-of-peak-paper>.

Roberts, Carys, Mathew Lawrence & Loren King, 'Managing automation: employment, inequality and ethics in the digital age', IPPR, 2017, <ippr.org/publications/managing-automation>.

Scheidel, Walter, *The Great Leveller: Violence and the history of inequality from the stone age to the twenty-first century*, Princeton University Press, 2017.

Scott, James C, *Seeing like a State: How certain schemes to improve the human condition have failed*, Yale University Press, New Haven, CT, 1998.

Segal, Lynne, *Radical Happiness: Moments of collective joy*, Verso, London, 2017.

Smith, Adam, *The Theory of Moral Sentiments*, Enhanced Media Publishing (Kindle), 2016 (1759).

Srnicek, Nick & Alex Williams, *Inventing the Future: Postcapitalism and a world without work*, Verso, London, 2015.

Tokumitsu, Miya, *Do What You Love: And other lies about success and happiness*, Regan Arts, 2015.

Tooze, Adam, *Statistics and the German State, 1900–1945: The Making of Modern Economic Knowledge*, Cambridge University Press, 2001.

Van Reybrouck, David, *Against Elections: The case for democracy*, Bodley Head, London, 2016.

Wall, Derek, *Elinor Ostrom's Rules for Radicals: Cooperative alternatives beyond markets and states*, Pluto Press, London, 2017.

Weber, Max, 'Politics as a vocation', 1919, <anthropos-lab.net/wp/wp-content/uploads/2011/12/Weber-Politics-as-a-Vocation.pdf>.

Weeks, Kathi, *The Problem with Work: Feminism, Marxism, antiwork politics, and postwork imaginaries*, Duke University Press, London, 2011.

Wilson, Katherine, *Tinkering: Australians reinvent DIY culture*, Monash University Publishing, Clayton, Vic., 2017.

Chapter opening quotes – sources

Introduction, p. 1

Le Guin, Ursula K, Foreword to *The Next Revolution: Popular Assemblies and the Promise of Direct Democracy* by Murray Bookchin, Penguin Random House, 2015.

Power, p. 19

Lyons, Tim, 'Is neoliberalism destroying social democracy?' Talk to the Australian Fabians by former ACTU assistant secretary, 26 April 2017.

Commons, p. 43

Malchik, Antonia, 'Who owns the earth?' *Aeon*, 12 April 2016, <https://aeon.co/essays/is-it-time-to-upend-the-idea-that-land-is-private-property>

Media, p. 73

Tancredi in *The Leopard* by Giuseppe Tomasi di Lampedusa, Pantheon Books, 2007

Government, p. 113

Greenleaf, Robert, *Servant Leadership: A Journey into the Nature of Legitimate Power & Greatness*, Paulist Press, 2002

Wealth, p. 140

Greenwald, Robert & Jesse Lava, 'Top Ten Koch Facts', *Huffington Post*, 4 October 2012, quote from billionaire businessman, Charles Koch, <https://www.huffingtonpost.com/robert-greenwald-and-jesse-lava/top-ten-koch-facts_b_1413499.html>

Work, p. 168

Russell, Bertrand, *In Praise of Idleness: and Other Essays*, Routledge, 1935, <https://archive.org/stream/in.ernet.dli.2015.218664/2015.218664.In-Praise_djvu.txt>

Education, p. 199

Arendt, Hannah, *On Revolution*, Penguin Books, 1963, <https://archive.org/stream/OnRevolution/ArendtOn-revolution_djvu.txt>

Joy, p. 232

King, Martin Luther, Jr.,'Where Do We Go From Here?' Annual Report delivered at the 11th Convention of the Southern Christian Leadership Conference, Atlanta, GA, 16 August 1967, <http://www-personal.umich.edu/~gmarkus/MLK_WhereDoWeGo.pdf>

ACKNOWLEDGMENTS

Books like this rely on the resources of a commons, the life in common, the intellectual and cultural information accrued by societies, added to and shaped by writers, researchers and teachers across time, and we all need to acknowledge that foundation. To paraphrase US senator Elizabeth Warren: you wrote a book? Good for you. But you wouldn't have been able to do it without a knowledge and information commons, built up over generations, and made available to you by various social institutions from schools to libraries to universities, and increasingly, by the shared spaces of the internet.

Along the way, there were some specific people who helped clarify my thinking, pointed out my errors, sharpened my approach, or made it possible in other ways for me to do what I do. I would like to thank Nicholas Gruen and John Quiggin, who both provided me with papers of theirs that helped immeasurably with the Government, Wealth and Education chapters. They are both generous contributors to public debate in Australia, and exactly what we mean we talk about the idea of public intellectuals.

I want to thank Jon Altman and Elise Klein who organised the two-day workshop on universal basic income in August 2017 and invited me along. They gathered a stellar selection of thinkers who, over those two days, pushed and prodded our approach to UBI and made us all think about it more clearly. In particular I would like to thank Frances Flanagan, Troy Henderson, Peter Whiteford, Eva Cox, Greg Marston and Jenny Mays. Tim Hollo also gave a brilliant paper at the workshop and has made himself available a number of times since to talk about matters arising, especially around the idea of the commons.

Speaking of the commons, I want to thank members of my circle of followers and followees on social media, whose writing and reflections provide a constant background conversation that helps alleviate the isolation of the lone writer. These people talk about important matters with style, grace, passion and – most importantly – humour, provide an endless stream of links and references that sends me scurrying down interesting rabbit holes, and they exemplify the best of what is possible with these new communication platforms. Many of them I know in real life, of course, but I want to acknowledge this online relationship in particular. Margaret Morgan, Graham Cameron, Kerryn Goldsworthy, Rob Schaap, Alison Croggon, Jim Parker, Helen Dale, Ian Millis, Terry Sedgwick, Anthony Georgeff, Matthew Clayfield, Amanda Rose, Shaun Cronin, Andie Fox, Andreas Ortmann, David Irving, Mark Bahnisch, Zoe Bowman, Renn Barker, James Bradley, Mark Davis, Jane Gilmore: thank you all and let's have coffee soon.

Nikki Lusk edited the book and made it much better. I am so grateful for her skill and professionalism. Alexandra Spring invited me to write for *The Guardian* about issues around the future of work, and is a generous and wise editor.

Finally, as ever, I want to thank my wife Tanya and my son Noah for their love and support. Tanya's range and depth of knowledge on public policy in general and education and work in particular means I have an incredible body of expertise to draw on and to check my thinking against. Noah makes me more proud every day as he pursues his career in dance with dedication, skill and love for the artform. They make every day better, every achievement more special, and they fill my life with joy.

INDEX